Potato Crop: Advances in Crop Science

Potato Crop: Advances in Crop Science

Editor: Patrick Garter

STATES
ACADEMIC PRESS
www.statesacademicpress.com

States Academic Press,
109 South 5th Street,
Brooklyn, NY 11249, USA

Visit us on the World Wide Web at:
www.statesacademicpress.com

ISBN: 978-1-63989-430-7 (Hardback)

Cataloging-in-Publication Data

Potato crop : advances in crop science / edited by Patrick Garter.
p. cm.
Includes bibliographical references and index.
ISBN 978-1-63989-430-7
1. Potatoes. 2. Crop science. 3. Potatoes--Breeding. 4. Potatoes--Diseases and pests. 5. Potatoes--Varieties. 6. Potatoes--Harvesting. I. Garter, Patrick.

SB211.P8 P68 2022
635.21--dc23

Table of Contents

Preface

Potatoes are starchy tubers of the plant Solanum tuberosum. It is a temperate crop which requires 300-500 mm of rainfall. Potatoes can grow upto 60 cm in height with its leaves dying back after flowering, fruiting and tuber formation. Their pollination is done by insects such as bumblebees along with self-fertilisation. Fruits and flowers of potatoes contain solanine while the tuber contains glycoalkaloids. Green potatoes have high amount of glycoalkaloids, making them unfit for consumption. Potatoes have more than 4000 varieties which are often differentiated by their waxiness and quantity of starch. The two varieties of starch found in potatoes are amylose and amylopectin. Other than food, potatoes are used extensively to produce alcohols, as livestock fodder and as thickener and binder for sauces and soups. This book brings forth some of the most innovative concepts and elucidates the unexplored aspects of potato crop. From theories to research to practical applications, case studies related to all contemporary topics of relevance to this field have been included herein. This book is a vital tool for all researching or studying potato crop as it gives incredible insights into emerging trends and concepts.

All of the data presented henceforth, was collaborated in the wake of recent advancements in the field. The aim of this book is to present the diversified developments from across the globe in a comprehensible manner. The opinions expressed in each chapter belong solely to the contributing authors. Their interpretations of the topics are the integral part of this book, which I have carefully compiled for a better understanding of the readers.

At the end, I would like to thank all those who dedicated their time and efforts for the successful completion of this book. I also wish to convey my gratitude towards my friends and family who supported me at every step.

Editor

1

Potato Breeding

Meredith W. Bonierbale, Walter R. Amoros, Elisa Salas, and Walter de Jong

Abstract The breeding of crop plants is a highly effective means of increasing agricultural productivity in a sustainable and environmentally safe way. Prebreeding and population improvement not only capture essential genetic resources and move desired traits along variety development pipelines but also help assure the creation of broad and dynamic gene pools to meet future, unanticipated needs. To efficiently meet multiple breeding objectives requires both interdisciplinary collaboration and a grasp of a wide range of scientific knowledge and expertise. This chapter addresses a range of topics that define and govern potato breeding, drawing from the experiences of both international and regional potato breeding programs, to orient readers to the interlinked components of population improvement and variety development. Using a case study approach to discuss breeding objectives together with respective implications for breeding needs, methods, and awareness-raising approaches for impact, we detail some key research and achievements contributing to current state of the art. Major populations under improvement at the International Potato Center along with breeding objectives and trait levels selected are described in terms of the agroecologies or uses they address in developing country national programs; these are contrasted with a discussion of the Cornell University program that is oriented to the northeastern US. A sample stage gate process, accelerated multi-trait selection schemes, heritability and heterosis exploitation, genomic selection, data management, and end user consultations are introduced in the contexts of these two programs. The topic of this chapter is supported and augmented with further details on subjects closely related to potato breeding, provided in chapters contributed to

M. W. Bonierbale
International Potato Center, Lima, Peru

Duquesa Business Centre, Manilva, Malaga, Spain

W. R. Amoros · E. Salas (✉)
International Potato Center, Lima, Peru
e-mail: CIP-Breeding1@cgiar.org; e.salas@cgiar.org

W. de Jong
Cornell University, Ithaca, NY, USA
e-mail: wsd2@cornell.edu

this volume by Ortiz and Mihovilovich, Ghislain and Douches, Burgos et al., and Ellis et al. The authors hope that the content serves to orient researchers and managers in countries with different degrees of development to plan and succeed in impactful potato improvement programs.

Implications of Genetics, Genepools, and Biology for Potato Breeding

Key Features

Commercial potato of world importance is a heterozygous, autotetraploid, clonal crop (2n = 4x = 48). Modern varieties are the products of extensive breeding between different cultivar groups and wild species. Potato varieties grown outside of South America since the end of the sixteenth century, as well as landrace (indigenous) cultivars grown in lowland Chile and in the high Andes are referred to as *Solanum tuberosum*, within which several groups are recognized (Spooner et al. 2014).

The genetic resources available for potato improvement comprise a polyploid series (2n = 2x = 24 to 2n = 6x = 72) with genetic features that facilitate gene transfer across ploidy levels. Farmers' landrace varieties dominate potato production in the Andean center of origin and diversity; as cultivars they offer tremendous trait diversity in readily useable form. Landrace cultivars, improved varieties and wild potato species in *Solanum* section *Petota* comprise GPs (gene pools) 1 and 2, according to the gene pool concept of Harlan and de Wet (1971) and are relatively straightforward to use in breeding. The wild tuber-bearing potatoes (section *Petota*) include some 200 species (see Spooner et al. 2014 for a review of taxonomic treatments). These typically produce only small tubers, which often contain high levels of toxic glycoalkaloids, and many require short days for tuberization. Domestication of wild potatoes for use as food likely involved selection for increased tuber size and reduced glycoalkaloid content, probably simultaneously in multiple locations in Andean and coastal regions of South America (Ugent et al. 1982, 1987).

Unreduced gametes, self-incompatibility, and inter-specific reproductive barriers played key roles in polyploidization and the maintenance of species boundaries during the evolution and domestication of potato, and knowledge of their genetics is useful in germplasm enhancement. Diploid potatoes are out-crossing due to a system of gametophytic self-incompatibility (Pushkarnath 1942; De Nettancourt 1977), which prevents inbreeding and thereby promotes intraspecific genetic variation. Tetraploid potato, on the other hand, is self-compatible. The breakdown of the gametophytic self-incompatibility system that operates in diploids is a common phenomenon in angiosperm polyploids (Frankel and Galun 1977; Levin 1983), but the molecular mechanism is not known (Comai 2005). Selfing results in severe inbreeding depression in most potatoes. Self-incompatibility at the diploid level, and inbreeding depression in both diploids and tetraploids, make it difficult to eliminate unfavorable alleles or drive favorable alleles to fixation. Several genes governing reproductive isolation, crossability, and ploidy are further discussed in the chapter contributed to this volume by Ortiz and Mihovilovic.

Potato varieties are maintained by clonal (vegetative, asexual) reproduction. Potato tubers are modified stems and comprise the vegetative "seed" used to propagate a variety. In potatoes with an even number of chromosomes (2x, 4x, 6x) it is possible to produce sexual (botanical) seed which provides for the generation of new genotypes. Once variation has been created in the form of sexual seed, any seedling has the potential to become a new variety via clonal propagation. Although most potatoes produce a large number of botanical seeds per fruit (~200) as well as many fruits per plant, clonal propagation results in a low propagation coefficient (five to tenfold per generation). Clonal breeding facilitates intentional and unintentional exposure of candidate varieties to pathogens. This helps breeders eliminate undesirable genotypes, but also necessitates steps to reduce exposure to detrimental viruses that are transmitted through vegetative (tuber) seed. In common with other root and tuber crops the limited ability to phenotype and potato for desirable morphological or developmental features is challenged by the underground location of the harvested product.

Genome Constitution and Variation

Maximum heterozygosity has been considered essential for performance of tetraploid potato, with inbreeding leading to reduced vigor and yield, flower bud abortion, lack of flower bud formation, and sterility (De Jong and Rowe 1971; Mendiburu and Peloquin 1977). However, empirical research with hybrid families has suggested that poor performance may be due to the expression of recessive alleles (De Jong and Rowe 1971) or that the presence of certain alleles may be more important for high yield (Bonierbale et al. 1993). Genomic studies have revealed that tri-allelic and tetra-allelic single nucleotide polymorphisms (SNPs) are rare in potato cultivars (Hirsch et al. 2013; 2014), though these might be expected in outstanding clones if yield or vigor were associated with maximum heterozygosity. Nevertheless, when the products of individual genes are amplified by PCR (polymerase chain reaction) it is not uncommon to find three alleles at a locus.

Xu et al. (2011) proposed heterozygosity as the key feature enabling the frequent occurrence of gene presence/absence variants and other potentially deleterious mutations in the genome of the heterozygous diploid potato clone "RH." Next-generation (short read length) sequencing of 807 genes from 83 potato cultivars revealed a tremendous amount of genetic variation in potato. On average, there is one variant (SNP or indel) every 24 base pair (bp) in exons, and one variant every 15 bp in introns (Uitdewilligen et al. 2013). The average minor allele frequency of a variant is low, though, at 0.14, and 61% of variants have minor allele frequencies less than 0.05 (Uitdewilligen et al. 2013). Given the sequence variation, and keeping in mind that tetraploid potato can contain up to four alleles at each locus, it is easy to understand why so much phenotypic variation results every time two heterozygous potato clones are crossed. Considerable effects due to dominance and epistasis are possible, and experience with such out-crossing polyploids has shown that the genetic variation due to dominance and epistatic effects is large compared to that seen in diploid crops (Gruneberg et al. 2009). This suggests that heterosis largely determines the performance of out-crossing, clonally propagated crops, although its basis in potato has not been fully elucidated.

The Cost of Increasing Genetic Variation

Most plant breeding training programs place heavy emphasis on the need to increase genetic variation. There is value in this, of course, as wild species contain many useful traits, genes/alleles not present in modern cultivars. But there is also a considerable cost that is not appreciated by those unfamiliar with potato breeding. Wild potato species contain countless alleles that are undesirable for potato production, and when a cultivated potato is crossed with a wild accession, oftentimes the offspring is quite poorly adapted. Many further generations of crossing are needed to eliminate the undesirable alleles, while keeping the desired ones. When embarking on such a venture it is important to realize upfront that the process typically takes decades. It may be possible to use molecular markers to speed the process up, by selecting against the donor genome (and for the gene of interest) after each cycle of crossing, but the process will still be much slower than adding a gene to cultivated potato by *Agrobacterium*-mediated transformation, or by editing an existing cultivated allele to a desired wild species allele with CRISPR-Cas9 (https://www.yourgenome.org/facts/what-is-crispr-cas9), once the respective target genes are known. It is worth noting that the "adaptation gap" between wild species germplasm and cultivated germplasm is growing over time, as breeders continually work to improve cultivated germplasm by increasing the frequency of desirable alleles and decreasing the frequency of undesirable ones. Despite the challenges, support for germplasm enhancement programs that strategically bridge this gap and enable the continuous influx of valuable genes from crop wild relatives through improved populations and into varieties is critical to meeting the world's growing need for food in the face of climate change.

Genetic Enhancement

Potato breeding strategies frequently include research to efficiently access traits from beyond the variety-ready germplasm base. Such trait research or pre-breeding is expected to result in new materials, methods, tools, knowledge, and approaches to support the breeding process rather than in finished products or varieties. Genetic and biochemical research often provides insights and tools that enable gains in traits toward new breeding objectives or improved program results. Pre-breeding is conducted in parallel with mainstream breeding activities, such as by developing support populations that are upgraded for trait levels or improved for agronomic traits so that new types of diversity can be introduced from un-adapted to adapted germplasm without impeding advance toward established breeding goals.

Before embarking on breeding for new traits or using uncharacterized germplasm, it is important to assess and consider positive and negative correlations among traits. Trait correlation influences the success of cross combinations and can determine breeding progress when multiple traits are concerned. Negative trait associations can be critical in achieving breeding progress and influence the choice of parents as well as the selection approach that will be most successful. Embarking on medium- to long-term population development does not result in new varieties after a single recombination and selection cycle, but should result in better parents that will help meet the medium- to long-term objectives of breeding programs.

Support populations are useful for enhancing diversity from un-adapted germplasm to avoid introducing undesirable features into advanced breeding populations. In the case of potato, wild and landrace relatives often carry undesired agro-morphological traits like deep eyes, small tuber size and late maturity, a requirement for short days, or high glycoalkaloid content, in addition to untapped resistance to biotic and abiotic factors and nutritional traits. Those undesired traits may be eliminated from hybrid populations by backcrossing to improved types, or source populations may be enhanced for agronomic traits before desired traits are transferred to improved genepools. The large majority of potato's genetic resources are diploid, and breeding at this level results in faster genetic gains than breeding at the tetraploid level. Thus, particularly for multigenic traits, pre-breeding in diploid source germplasm before incorporating new traits into tetraploid breeding populations can be very effective.

Introgressing novel traits from distant wild species such as those in the tertiary gene pool (GP3) may require the use of bridge species to circumvent interspecific reproductive barriers, and unreduced gametes to transfer traits across ploidy levels. When interspecific crossing is possible, backcross schemes are modified from those used for inbred crops, such that a different genotype of the recipient germplasm (adapted type) is used in each cycle of crossing with the trait donor or selected hybrid to avoid inbreeding depression. This is illustrated by Gaiero et al. (2017) who introgressed partial resistance to bacterial wilt, caused by *Ralstonia solanacearum*, from the sexually incompatible GP3 species *S. commersonii* by bridge crosses with 2x *S. tuberosum* Phureja Group and successive backcrosses with different *S. tuberosum* Group Tuberosum genotypes. The resulting advanced backcross progenies are now being used by several breeding programs due to the acute need for resistance to bacterial wilt.

Case Study 1: Genetic Enhancement and Incorporation of Iron Content from Diploid into Tetraploid Cultivated Potatoes

With support from HarvestPlus (www.harvestplus.org) advocating the breeding of staple crops for micronutrient density, CIP has sought to increase the content of iron and zinc in potato through an inter-ploid breeding strategy. Recurrent selection in a base population of landrace potatoes took advantage of greater response to selection at the diploid level, and resulted in a population of diploid potato reaching 35 ppm iron and 30 ppm zinc from a baseline of 20 and 16 ppm, respectively. Elevated iron and zinc contents achieved in the source germplasm were incorporated into advanced, tetraploid populations via unreduced gametes. The resulting tetraploid potato population had iron and zinc concentrations twofold higher than baseline levels, which approaches the breeding targets for human populations with deficien-cies of these minerals and high potato intake. Further work remains to be done to identify clones adapted to target agro-ecologies that maintain as favorable a package of traits as possible, including new levels of iron or zinc concentrations in resilient, consumer-accepted table potatoes.

Analyzing biofortified populations under recurrent selection in diploid potato germplasm, Paget et al. (2014) found moderate to high and positive correlation between iron (Fe) and zinc (Zn) contents from cycle I ($r^2 = 0.45$) and from cycle II ($r^2 = 0.72$), indicating that evaluation and selection for one of these traits will result in concomitant increase in the other. Negative genetic correlations were found between dry matter and Fe, Zn, Ca, and vitamin C contents (genetic correlation close to zero for vitamin C in Cycle 1) when analyzed on a dry-weight basis. In contrast, the same genetic correlation estimates were positive (but small) when analyzed on a fresh-weight basis.

The genetic correlations in this example are strong enough that under multi-trait selection the breeding population is improved simultaneously for iron and zinc concentrations, whereas the average tuber weight of the population decreases, and the number of tubers per plant increases. A reduction

Table Additive genetic correlations (are in bold) and confident intervals (are in brackets) for Fe, Zn, Ca, vitamin C, and tuber dry matter content from a multivariate analysis of Cycle1 data (Model 4) estimated on a dry and fresh-weight basis

Trait	Iron	Zinc	Calcium	Vitamin C
Cycle I (dry weight basis)				
Zinc	**0.45** [0.32 0.64]			
Calcium	**0.04** [−0.23 0.34]	**0.12** [−0.15 0.39]		
Vitamin C	**−0.01** [−0.18 0.29]	**0.10** [−0.15 0.30]	**0.05** [−0.27 0.33]	
Dry matter	**−0.23** [−0.42 −0.06]	**−0.24** [−0.41 −0.07]	**−0.19** [−0.36 0.07]	**−0.06** [−0.28 0.10]
Cycle II (dry weight basis)				
Zinc	**0.72** [0.42 0.88]			
Calcium	**0.35** [−0.04 0.61]	**0.57** [0.18 0.76]		
Dry matter	**−0.34** [−0.61 0.08]	**−0.38** [−0.66 0.10]	**−0.14** [−0.49 0.20]	
Cycle II (fresh weight basis)				
Zinc	**0.61** [0.33 0.84]			
Calcium	**0.07** [−0.32 0.52]	**0.45** [−0.02 0.77]		
Dry matter	**0.18** [−0.13 0.36]	**0.14** [−0.13 0.38]	**0.05** [−0.23 0.27]	

From: Paget et al. (2014)

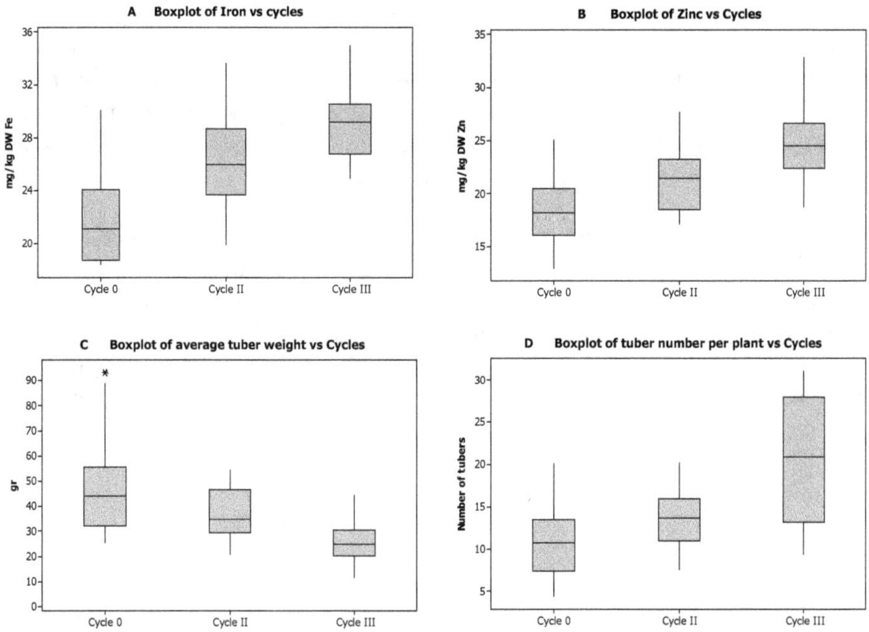

Fig. Box plots of ranges of trait variation through three cycles of recurrent selection in diploid potato germplasm. (**a**) Iron concentration; (**b**) zinc concentration; (**c**) average tuber weight (reflecting size); (**d**) Tuber number per plant

in tuber size was encountered during population improvement at the diploid level even though it was considered in selection. This loss was mitigated, however, by returning to the tetraploid level via 4x–2x crosses by which the gains in the mineral contents realized at the 2x level were transferred to advanced populations.

Principles of Potato Breeding Methods and Approaches

The principal advantage of breeding clonally propagated crops is that each clonal variety is fixed and simple to maintain. Genetic purity is less of an issue in vegetatively—than in sexually—propagated crops. One substantial disadvantage of vegetative propagation though is that diseases are easily transmitted across clonal generations during propagation; another is that potato planting material is bulky and perishable by nature and the production of healthy material is expensive.

The single most challenging aspect of potato breeding is the identification of superior individuals that combine as many high priority traits, and as few weaknesses, as possible from a given cohort of F1 progeny in a reasonable time frame. Additional important challenges include the improvement of support populations, and the selection or construction of parents. The parents used for crossing are highly heterozygous—exceptions are inbreeding lines generated by self-fertilization or doubled-monoploid production. Each potato seed that results from a heterozygous × heterozygous cross differs, at many loci, from any other seed from the same cross. The resulting heterozygous genotypes are subject to selection after being fixed (stabilized in genetic terms) by clonal propagation.

Polyploidy, heterozygosity, and heterosis make the identification of good parents particularly challenging. At present, the performance of a parent can only be determined after the fact, that is, by looking at its progeny. Advances in genomic selection may make it possible to identify good parents in advance. Heterozygous tetraploid potato genotypes harbor great allelic diversity and interactions that are responsible for their performance as clones. As parental clones are not inbred, a genotype can never be reconstituted after sexual crossing or self-pollination. A practical consequence of this for potato breeding is that many, many, traits segregate in the progeny of any parent or parental combination. Directional breeding results in incremental changes in gene frequencies among progenies.

In the course of potato breeding an "F1 clone hybrid" is generally crossed with another "F1 clone hybrid," and the progeny is heterogeneous and has an extreme large segregation variance. A good parent generates a large genetic variation around a high family mean for a given trait. Once heterogeneous and heterozygous progenies are generated by crossing two potato clones, selection of clones within a given pool of genetic variation for variety development is conceptually, if not technically, straightforward. All the genetic advantages of clonally propagated crops can be used for variety development, and the genotype that will finally be released is among the progeny immediately after the initial crossing.

Inter-group crosses are important in population improvement of clonally propagated crops. Gruneberg et al. (2009) have suggested that this aspect of clone breeding is often neglected and may be the reason for low breeding progress in many clonally propagated crops when compared to the improvement of sexually reproducing ones. Complementary germplasm groups have not been identified in potato but strategies to maximize heterozygosity have been proposed to increase yield in tetraploid potatoes (e.g., Chase 1963). However, it is unlikely that the direct relationship of maximum heterozygosity and yield will extend to crosses involving unadapted germplasm (Bonierbale et al. 1993). Thresholds for heterosis were suggested upon finding of increased yield and vigor in two-way hybrids (cultivated × wild species) with respect to crosses within the cultivated genepool, but no additional increments in hybrids involving cultivated potato and two wild species (3-way hybrids) (Sanford and Hanneman 1982). Thus, it is unlikely that heterotic groups could be established in potato germplasm on the basis of genetic distance measures or maximum heterozygosity alone.

Population Improvement

Medium- to long-term genetic gain can be achieved over sequential cycles of crossing and selection. Recurrent selection is defined as reselection generation after generation, with inter mating of selected plants to produce the recombinant population for the next cycle of selection. The goal of recurrent selection is to improve the mean performance of a population of plants; a secondary goal, but nevertheless also important, is to maintain as much genetic variability as possible. Open recurrent selection is a method for improving the mean performance of a population while maintaining and increasing genetic variability by periodically introducing new sources of traits under selection. In genetic enhancement a few cycles of recombination facilitate the breakdown of linkage blocks so that desired traits from unadapted germplasm can be carried forward and undesirable ones left behind, minimizing the effects of linkage drag thus shortening the time required to introgress or incorporate new diversity into advanced populations. Depending on the stage of a population under improvement, selections may be considered for use as varieties, or as parents that will contribute to further gains toward complex breeding objectives. Figure

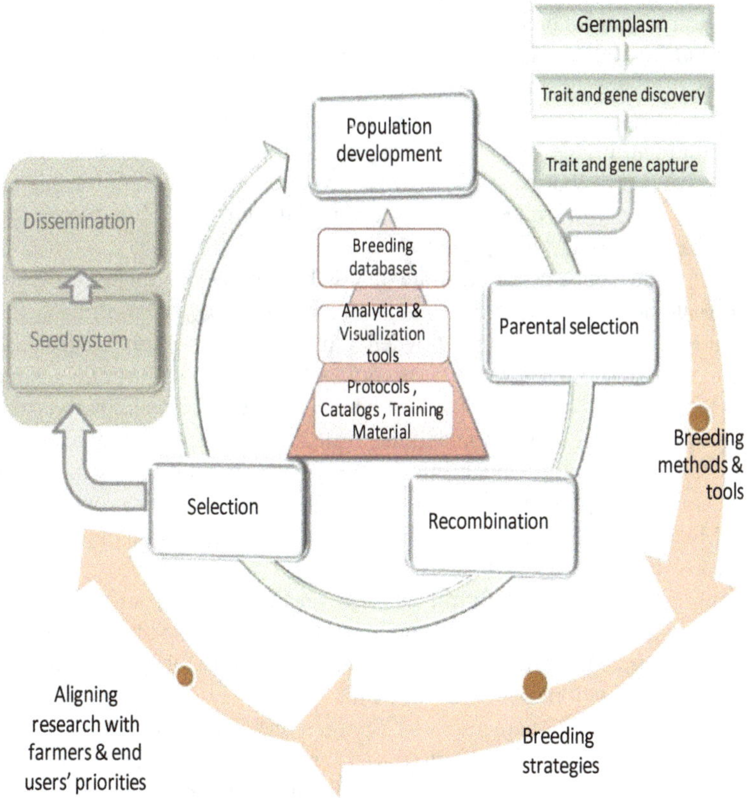

Fig. Schematic representation of components of a potato breeding program. (Courtesy of Elisa Mihovilovich)

depicts simultaneous and integrated practices of recurrent selection, genetic enhancement, and variety selection in a potato breeding program.

In polyploid potato, more than one allele per locus can be transferred to the next generation in gametes, and thus, in contrast to diploids, the genetic variation due to dominance influences the response to selection in population improvement as long as the population is not in equilibrium. In tetraploid potato, a population is usually not in equilibrium after recombining parental material in controlled crossings, and 1/3 of the dominance variance is exploitable for selection progress when selection takes place on the female and male side [for further details see Wricke and Weber (1986) and Gallais (2004). The exploitation of the dominance variance in population improvement, in combination with faster genetic gains in diploids than in tetraploids, and the inheritance of 2n gametes as described by Ortiz (1998) provide great diversity and flexibility for potato improvement.

Crossing Parents

The choice of parents is an important step in any breeding program, especially for medium- and long-term breeding progress. The identification and number of crosses to perform at a given breeding stage is a factor of the knowledge available on potential parents (i.e., trait constitution and breeding value), the breeding objective, and facilities and resources available for crossing and selection. As mentioned above, the polyploid and highly heterozygous nature of potato mean that dominance and epistatic effects contribute considerably to clone performance, and for this reason, little is known about the value of a parent or specific cross combination until it has been tested.

Parental value can be assessed strategically through the conduct of progeny tests in appropriate breeding designs, the evaluation of pedigrees, or empirically through observation of selection ratios. Even when hundreds of crosses are made, it is often observed that the best clones trace back to very few crosses. Thus, it is desirable to predict which genotypes will be the best parents, since if this is known, efforts can be concentrated on the generation and the evaluation of the most promising combinations. Regardless of how much information is available to support the choice of parents, a practical approach is to sow around 200 seeds of every cross to be evaluated, and then sow more seed or emphasize the parents of those combinations that performed well in future years.

Parental value is a function of the genetic constitution of a trait donor. Complex traits are often comprised of several components, and defining those that contribute reliably to influence trait values is a prerequisite to successful identification of parents. Accurate evaluation of progeny is the most effective means to identify superior progenitors of inherited characteristics. Sprague and Tatum (1941) introduced the concepts of general combining ability (GCA) and specific combining ability (SCA) to distinguish between the average performance of parents in crosses (GCA) and the deviation of individual crosses from the average of crosses (SCA).

Studies of combining ability make it possible to identify parents (those with good GCA) that will perform well in most crosses, and also facilitate the development of superior hybrids through the use of parents with high SCA. Genetic analyses to measure combining ability further allow breeders to estimate genetic effects and parameters such as trait variance, covariance, correlations among traits, heritability, and the relative importance of additive and dominance variation. GCA represents mainly additive and additive × additive types of genetic variance. Thus, when a potato clone is selected as having good GCA, it means that the algebraic sum of the additive and additive × additive epistasis effects it passed on to its offspring produced a favorable result in excess of the average of all the offspring tested. GCA has a conceptual implication that each line being evaluated is tested against a large random sample of some specified population (Plaisted et al. 1962). In actual tests for GCA, however, the testers (usually males) often represent only a limited sample of a population.

Mating Designs

The term "mating design" refers to the mating of parents in a systematic plan of crosses to determine genetic parameters and/or parental value. These procedures are particularly useful for the identification of plants that will be the most effective parents in pre-breeding or recurrent selection programs. However, systematic matings can be difficult to achieve because of sterility and incompatibility encountered during crossing. Mating designs have been classified according to the number of factors to be analyzed, the parents, and modalities of combination. Some mating designs are used more extensively than others, but each has its advantages and disadvantages depending on the reference population under consideration and the information desired. Following are descriptions of some of the most important designs for identifying superior progenitors to help assure genetic gains in a potato breeding program. Figure, illustrates some of these mating designs.

Diallel Analysis is the best way to determine the combining ability of parents. It consists of the analysis of a set of crosses produced involving "*n*" lines in all possible combinations, a so-called diallel cross. The diallel mating design has been used extensively in potato germplasm enhancement and can be very useful if properly analyzed and interpreted. The analysis of diallels provides information on GCA and SCA of parents and their crosses and makes it possible to determine if reciprocal crosses give equivalent results.

The most commonly used *methods of diallel designs* are those proposed by:

- Griffing (1956), by which general and specific combining abilities are estimated;
- Gardner and Heberhart (1966), by which the variety and heterosis are evaluated; and
- Hayman (1954), in which information regarding additive and dominance effects for a characteristic and the genetic values of the parents is used.

Fig. Schematic representation of different mating designs in which parent lines are indicated with numbers, and the cross combinations between them are represented with checkmarks in shaded cells. (**a**) Partial dialel, (**b**) Line × tester, and (**c**) Design II

A

Partial diallel									
		Males							
		1	2	3	4	5	6	7	8
Females	1				√	√	√		
	2					√	√	√	
	3						√	√	√
	4							√	√
	5								√
	6								
	7								
	8								

B

Line x Tester				
		Testers		
		1	2	3
Lines	1	√	√	√
	2	√	√	√
	3	√	√	√
	4	√	√	√
	5	√	√	√
	6	√	√	√
	7	√	√	√

C

Design II					
		Males			
		5	6	7	8
Females	1	√	√	√	√
	2	√	√	√	√
	3	√	√	√	√
	4	√	√	√	√

One of the most used is *Griffing's Method 2*, which estimates GCA and SCA) relating mainly to additive and non-additive gene effects (dominance and epistasis).

Partial Diallel

Analyses involving "*n*" parents in all possible combinations become unmanageable as the number of lines (*n*) increases. On the other hand, if only a small number of parents are tested, the estimates of combining ability tend to have a large sampling error. These difficulties led to the development of sampling crosses produced by large numbers of parents, without affecting the efficiency of the diallel technique. In a normal diallel, each line is involved in $(n - 1)$ crosses. Kempthorne and Curnow (1961) presented the concept of the partial diallel design in which only a random sample of crosses, say of size'*s*, is analyzed where "*s*" is less than $n - 1$.

Line × Tester Design

The Line × Tester Design also provides information about the GCA and SCA of parents, and is also helpful in estimating various types of gene effects. The *crossing plan* of this design is as follows:

"*l*" lines are crossed to "*t*" testers so that $l \times t$ full-sib progenies are produced;

These progenies, with or without parents (i.e., testers and lines), are then tested in a replicated trial using a suitable experimental design, say randomized block design.

If there are three testers and seven lines, there are $7 \times 3 = 21$ crosses. For evaluation, the 21 crosses along with 10 parents, for a total of 31 entries, might be tested in a randomized complete block design with four replications. For this case, uniform planting material must be produced to enable inclusion of parental clones and progeny in the same trial.

Design II

In mating Design II (or Factorial Design), described by Comstock and Robinson (1948), the genetic information is similar to that obtained with Diallel Analysis.

Different sets of parents are used as males and females. If a set of eight parents is included in the design II, 16 crosses will be obtained. This design is advantageous when not all clones to be tested are male or female fertile.

Breeding Values

The estimated breeding value (EBV) of an individual can be calculated on the basis of pedigrees and performance in the course of a breeding program, and does not rely on the conduct of mating designs. Using appropriate statistical analysis, breeding value predicts how useful each individual would be as a parent; it expresses the ability of a parent to pass on superior trait levels to its offspring and is used for ranking breeding performance of an individual as a parent relative to the population average. The use of information on the individual and all relatives greatly increases the accuracy of selection in a breeding program (Lynch and Walsh 1998).

The calculation of breeding value goes beyond the typical estimation of genotypic or parental values with models based on fixed effects, by enabling the estimation of random effects of a mixed model. Mixed linear models are able to model different covariance structures and thus provide an improved representation of the underlying random and error components of variance (Oakey et al. 2007). The application of mixed models to estimate breeding values uses pedigree information to model and exploit genetic correlation among relatives and applies flexible variance–covariance structures for genotype-by-environment interaction to accurately predict performance (Piepho et al. 2008). The improved accuracy afforded by the use of all data in a breeding program with mixed models allows the analysis of repeated measures, unbalanced design experiments, spatial data, and multi-environment trials.

Breeding value has been successfully applied in several crops (sugarcane, eucalyptus, soybean, maize, and even potato). Slater et al. (2014) illustrate selection in potato based on breeding values. They conclude that using best linear unbiased prediction (BLUP) and pedigree to estimate breeding values can result in increased genetic gains for low heritability traits in auto tetraploid potato. Theory and applicability of breeding values in quantitative trait improvement are illustrated in Bernardo (2002) Relevant statistical packages that fit linear mixed models to large data sets using the Residual Maximum Likelihood (REML), approach, such as the Asreml-R reference manual (Butler et al. 2009) provide powerful software for the use of breeding values in plant breeding.

In animal breeding, EBVs are the basis for marketing breeding parents and they provide breeders with critical information for selection decisions. In applying EBVs, it is important to achieve a balance between the different groups of traits and to place emphasis on those traits that are important to the objective population, markets, and environment. It is not feasible, nor always desirable to seek high EBVs for all traits in a single progenitor. In fact, a comprehensive range of EBVs has the advantage that it is possible to avoid extremes in particular traits and select for animals/genotypes with balanced overall performance. The method is particularly useful in non-inbred populations and potato breeders should pay careful attention to this analytical approach.

Early Versus Late Generation Selection

In early generations it is possible to select for highly heritable traits for which accurate assessment of a genotype can be carried out on one or a few plants, but for more genetically complex traits, and for traits where interplant competition is an issue, it is necessary to evaluate clones in multi-plant plots planted as blocks under homogenous field conditions. The aim is an unbiased comparison of genotypes within blocks. The number of plants per plot and the number of replications or blocks depends on the breeding stage. The low propagation coefficient of potato (about ten depending on the propagation method used) limits the amount of planting material available at each stage of selection and is one reason that potato breeding is relatively slow (at least 8–10 years and usually more, from crossing until variety release). Large numbers of genotypes can be assessed for simple traits in early stages when small amounts of seeds are available; while more complex traits are assessed in later stages when larger quantities of seeds are needed for replicated trials, though less genetic diversity is represented.

Multistage selection can be managed in subsequent steps from early to later stages, or with indices that use a weight assigned to each trait. In practice, some characters are selected sequentially especially where there is clearly a lowest acceptable value (tuber size, shape, and color as well as pest and disease resistance), while others are selected simultaneously by aggregating characters into an index (often an intuitively formed index such as score values of overall performance.

Case Study 2: The Use of a Selection Index in Potato Breeding

An early-generation (seedling stage) selection method was applied at CIP for the identification of families and individuals that tuberize well under long days and high temperatures. Five groups of families generated by intercrossing CIP's best advanced parents from two populations and long day-adapted varieties underwent a greenhouse test during summer, applying two photoperiods: natural short days of 12 h and simulated long days using lamps to extend the photoperiod to 16 h. Four plant morphology prototypes were identified according to patterns of above- and below-ground growth including branching, stolon, and tuber formation.
Individual plants of each family were evaluated at harvest taking into account parameters including breeder's preference, a tuberization score based on the four patterns, tuber uniformity and physiological disorders including sprouting tubers, knobbiness, chain tubers, and cracking.

A selection index (SI) was built assigning a weight to each of these parameters, three for breeders' preference (BP), two for tuberization score (TS) and tuber uniformity (TU), and one for physiological disorders: sprouting (Sp), knobbiness (Kn), chained tubers (Ch), and cracks (Cr)

$$SI_n = BP_n(3) + TS_n(2) + TU_n(2) + Sp_n + Kn_n + Ch_n + Cr_n$$

Families with a high frequency of progenies exceeding an estimated selection index of 54 were selected. Seven of these selected families and three unselected were used in validation studies under field conditions.

Fig. Plant morphology and growth patterns observed among families in early generation screening for adaptation to warm, long day environments. (**a**) Typical morphology under 12:12 h light:dark regime; (**b**) Typical morphology under 16:8 h light:dark regime; (**c**) Above-ground morphology of representative plants with growth patterns I, II, III, and IV; Below-ground morphology of representative plants with growth patterns I, II, III, and IV

Stability and Adaptation

Breeding programs typically breed for several locations rather than just one. Hence field evaluations that underpin selection must simulate a range of environments. For this reason, within limits of the propagation coefficient, clones are tested in plots within homogenous blocks at several locations over several years. The objective of multilocation trials is to assess promising genotypes with respect to narrow and broad adaptation. In practical potato breeding, "broad adaptation," the ability to perform well in a range of environments, appears essential for a clone to have any meaningful commercial success.

The wide range of quality preferences and the numerous pests and diseases addressed by potato breeding programs dictates the exposures required for selection. Decentralized selection is generally required by a program with multiple target environments and is best designed to address sets of traits that can logistically be assessed together in target environments or sites that represent them well. Along the course of selection, a combination of on-station and on-farm trials are usually performed.

National programs requesting potato breeding materials from CIP are provided with germplasm from different selection stages depending on their capacity to evaluate clones and the relative suitability of advanced germplasm to their production targets. Advanced clones and parents are available as in vitro plantlets. The genotypes provided meet multiple selection criteria for the given target environment, as assessed in similar testing sites. Upon receipt they are propagated and subject to multilocational trials before use as parents in national programs or testing as varieties.

Case Study 3: Breeding and Variety Development in Bangladesh

Breeding for stable yields and multiple disease resistance at CIP takes advantage of germplasm collections, broad-based advanced populations, and environmental diversity in Peru. One such population is the advanced lowland tropics virus resistant population, which is adapted to dry arid regions where virus pressure is high. With support from GIZ/BMZ and follow-up support from USAID, CIP provided 35 potato breeding lines of this population as in vitro plants to the Tuber Crop Research Centre of the Bangladesh Agricultural Research Institute in 2009. After local evaluation over a period of 7 years, a salt tolerant variety (BARI Alu-72) and a heat tolerant variety (BARI Alu-73) adapted to Bangladesh were released.

When distant environments can only be partially simulated in CIP's breeding program, samples of true seed families (TSF) from good parents can be distributed. TSF are available for national programs with the capacity to carry out early generation selection as well as variety identification trials. One advantage in working with TSF is the considerable reduction in time-to-release compared to that required when receiving advanced clones. Conducting selection with true seed in target environments can save 6 years or more, i.e., time that would otherwise be required for CIP breeders to conduct selection from the same TSF at their trial sites and recommend elite clones as in vitro plants.

When national programs test CIP germplasm as TSF, valuable information is returned to CIP regarding family performance (i.e., frequency of selected clones in each family), and this serves to refine the concept of adaptation and best bet materials for that location. National programs that invest in identifying parents based on local performance, and then utilize these parents for crossing and recurrent selection, benefit from knowing which breeding materials will serve those best and are likely to experience significantly faster genetic gains over the medium to long term.

Case Study 4: Collaborative Breeding in Vietnam

Advanced germplasm from CIP's lowland tropic virus resistant (LTVR) and late blight heat tolerant (LBHT) breeding populations developed in Peru were dispatched to Vietnam in the form of in vitro plants from 2005 to 2010. The materials were used in two ways: (i) for direct variety testing, and (ii) in cross-breeding to improve adaptation to local conditions.

Results of the collaboration demonstrated both the direct utility of clones from Peru in Vietnam—through the identification of two outstanding clones recommended for variety testing, and the advantage of local breeding, yielding, after three rapid cycles of recurrent selection, a new generation of elite clones with higher yields than the elite clones provided from Peru. The yield increase from local breeding and selection was realized in a considerably less time (9 years) than would have been required for the development, introduction, and testing of a new cycle of elite clones from CIP (at least 12 years).

There are several advantages of testing early generation potato progenies at more than one location. Information from contrasting environments can be combined if the breeder tests a clone at two or more locations; those few clones that are broadly adapted can be identified early in the selection process. In addition, if one location experiences low disease pressure, daughter tubers from that site can be used to provide clean planting material for the next generation of selection. The healthy seed plot provides for an assessment of performance in the absence of major production constraints, while various pressures can be applied at additional sites. Alternatively, a protected environment (e.g., quarantine screen house) can be used to maintain a healthy copy of each selected genotype, but unlike the maintenance of a healthy

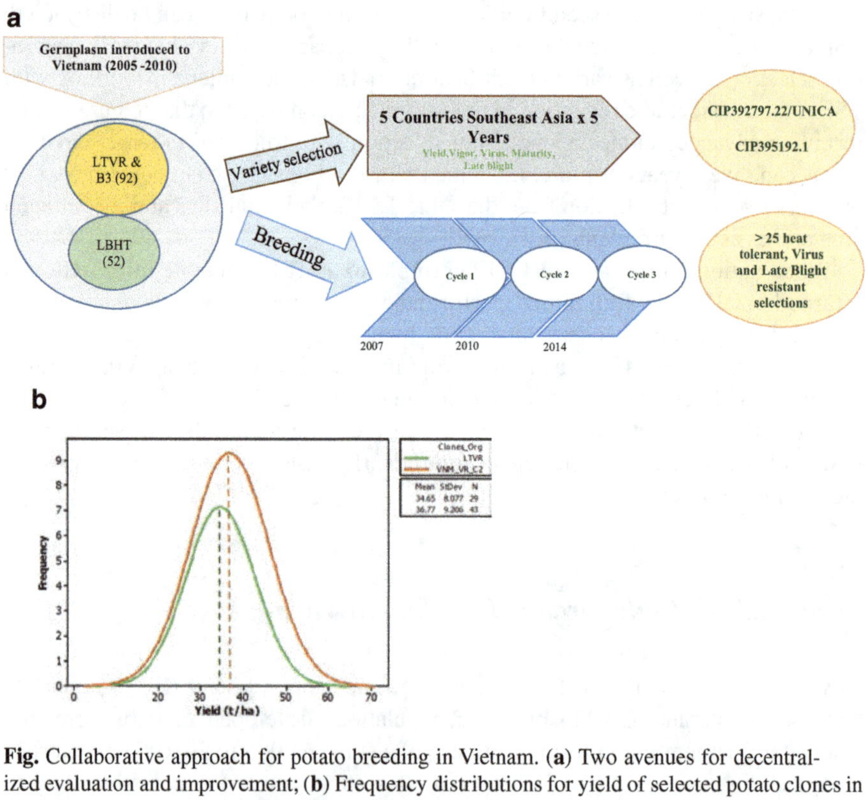

Fig. Collaborative approach for potato breeding in Vietnam. (**a**) Two avenues for decentralized evaluation and improvement; (**b**) Frequency distributions for yield of selected potato clones in three cycles of recurrent selection conducted in Vietnam

seed plot, this typically does not allow for significant production of seed to support evaluation along the breeding cycle.

Rapid and real-time exchange of information on selection percentage between breeders and selectors helps inform breeders on the future crossing plans most likely to meet the requestor's needs. This simple data can confirm the breeders' concept of GCA and suitability of test environments with respect to the targeted ones. Breeders can use the information returned from selectors regarding percentage of selection in controlled cross families to exploit heterosis through cross combinations. Selection of parents and cross prediction can probably be improved by systematic collection of data on the percent of clones selected in families by the assignment of parents with good GCA or mutual SCA to complementary gene pools. This would benefit both CIP's and the national programs' breeding efforts, if it resulted in increased frequency of high-yielding clones with needed traits, or potentially, in yield jumps.

Importance and Relation of GCA and Heterosis in Potato

Heterosis was examined in a new tetraploid potato hybrid population obtained by crossing members of two advanced breeding populations developed at CIP for adaptation to the highlands (B3) and subtropical lowlands (LTVR). Significant positive heterotic values ranged from 18 to 60% for mid parent heterosis and from 16 to 162% for best parent heterosis or heterobeltiosis. Positive heterotic values were found for tuber yield and tuber number in each environment. Significant GxE interaction on heterosis showed a differential effect of the environment on the magnitude of heterosis expression.

The expression of heterobeltiosis within B3 and LTVR confirms the effectiveness of the population breeding method on maintaining a broad genetic base (Mendoza and Sawyer 1985). Mid-parent heterosis can be the result of the combination of good levels of GCA of parents and some level of SCA of the cross. A reciprocal recurrent selection scheme (RRS) has been initiated to benefit from intercrossing between best parents of two complementary populations, B3 and LTVR (RRS), also known as recurrent reciprocal half sib selection, is a form of recurrent selection used to improve both GCA and SCA of a population for a character or characters.

Potato Breeding Procedures Overview

Hybridization

Heterozygous parents are recombined in controlled biparental combinations or by use of bulk pollen to create new variation in the form of sexual seed. The crossing block is established after selecting the parents to be used as females and as males, and the "planting on a brick method" or modifications of it can be practiced to encourage flowering and fruit set over a period of about 10 weeks. In the "brick method," potato tubers are placed on bricks partially buried in the soil. The tubers sprout and stems are allowed to grow. After some time the soil around the base of the full-size plant is washed away with a spray of water, which exposes emerging stolons and small tubers, but does not disrupt the root system that has penetrated in and around the brick. Thereafter, stolons and tuber that emerge from the base of the plant are removed. Potato has a complete flower with five anthers and a stigma that becomes receptive as pollen is shed from open flowers. Emasculation is practiced to prevent selfing when genetic studies will be performed, when parental value will be assessed in progeny tests and for best control of pedigrees (record-keeping).

Crossing at CIP is enhanced by flowering induction. Under long day conditions most potato clones flower to some extent. Typically, *S. tuberosum Phureja* Group flowers under both long and short-day photoperiods. Clones of Group *tuberosum*, however, usually will not flower under short days. In addition, there are many clones, especially those of early maturity, that rarely flower under any conditions or that flower sparsely and over a very short time period. For these reasons it is often desirable to induce flowering. This can be done by either of two methods: "grafting" or "planting on a brick." In the grafting method shoots of potato are grafted onto the stems of tomato plants. In both methods the idea is to prevent potato from developing beyond the vegetative flowering phase to the tuberization-senescense phase. In addition to these mechanical methods, other traditional flowering-enhancing practices may be followed such as long-day lighting, sprays of 40 ppm gibberellic acid at 4-day intervals, and heavy nitrogen fertilization. Temperature and humidity are also important. Ideal temperatures are 20 °C day and 16 °C night. Humidity should be 80% or higher. Procedures and crossing techniques for potato breeding as practiced at CIP are documented at https://research.cip.cgiar.org/potatoknowledge/proceduretechniques.php. A modification of "planting on a brick" using peat pots is shown. The open source software CIPCROSS can be used to document all aspects of the crossing block, including pedigrees, storage, and inven-tory using bar coding for accuracy and efficiency.

Selection Schemes

Selection of potato varieties from botanical seed is conducted in several steps. Figure shows how the crossing of two parents might be followed by five sequential steps in time (one selection step conducted with seedling plants and four subsequent selection steps conducted with cloned plants from tubers of the seedling genotypes). This scheme is an oversimplification since many crosses or families are developed simultaneously and many seeds of each family are subject to evaluation and selection at the same time, while here only three seedlings in each of 11 families are shown.

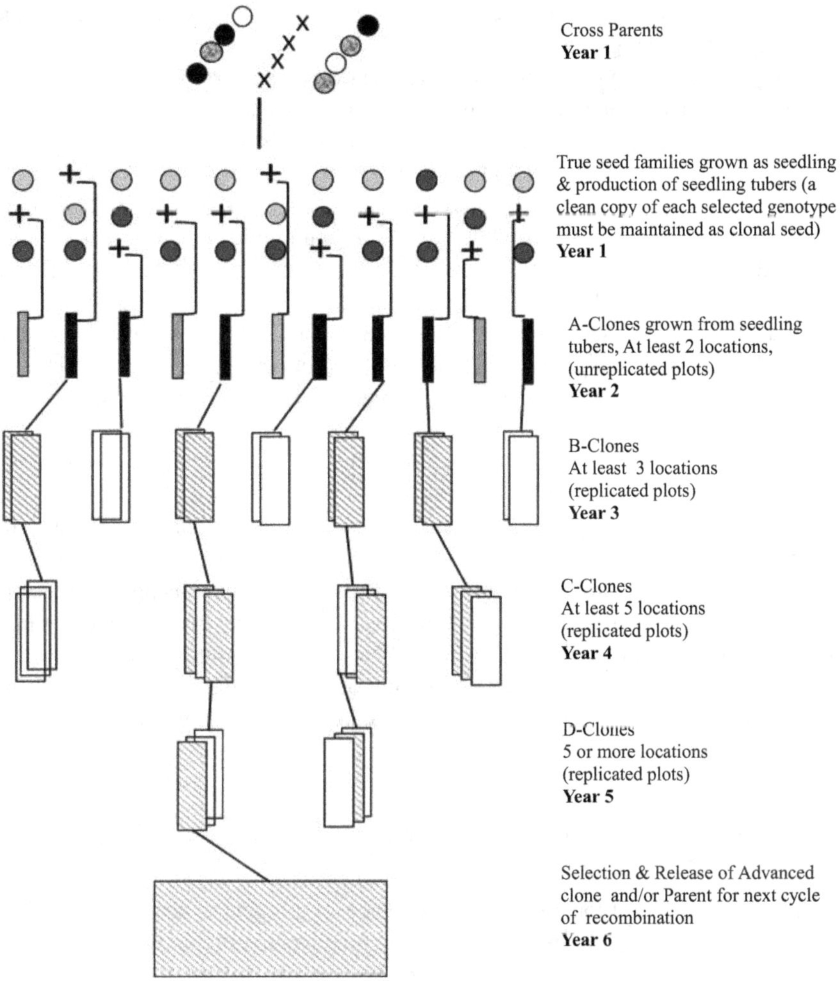

Fig. Multilocation Clonal selection Scheme. (Modified from Gruneberg et al. 2009)

In principle, if the true seed plant could be cloned in large quantities and the population could be assessed with adequate accuracy, it is theoretically possible to select the "best" genotype in the first year, since maximum variation is present at the start of each selection cycle and no genetic changes are realized during it. In practice though, the individual genotypes selected in early stages of each clonal selection cycle are subject to propagation via tubers to permit adequate plot size and replication for later-stage assessments.

The overall selection scheme should consider the vulnerability of clonal planting material to disease. At least some tubers of every clone should be planted in a disease-free location, or time and resources should be allocated for the eventual elimination of pathogens from successful clones before they can be distributed among geographic regions.

Many breeding programs use only one location at the early selection stage, preferring to evaluate in favorable conditions such as on station, while also producing healthy seed for use in later selection stages. This location should be as free from virus pressure as possible to help keep seed clean over the years. For example, the New York breeding program makes all evaluations for the first 2 clonal years in a single location, going to a second location (for an unreplicated evaluation of yield) in the third clonal generation, and multiple locations (for replicated yield trials) starting in the fourth clonal year. Typical numbers of genotypes per cycle in this program are: 20,000 seedlings in year 1, 18,000 four hill plots in year 2, 1500 twenty-hill plots in year 3, and 250 hundred-hill plots in year 4.

Nevertheless, the first clonal generation can be evaluated at more than one location if sufficient clonal seed can be produced. For simultaneous evaluation at up to three locations in the first clonal generation, true seed families (TSF) are first converted to tuber families (TF) so that at least four copies of each genotype from each family are produced. Methods for evaluating TF include (1) planting three to five tubers of each genotype in a single row at a single location or (2) equivalent samples of each tuber family are evaluated in two or more locations, using one tuber per genotype and location. Multilocational testing of first clonal generation tuber families can be done with, or without the identification and labeling of individual genotypes. Detailed procedures for the production of tuber families followed at CIP are illustrated at: https://research.cip.cgiar.org/potatoknowledge/tuberfamilies.php

Data Management and Analytical Tools

Standardized information on the performance of progenies and selected clones across environments is necessary in order for breeders to efficiently make decisions about selection and variety release. The methods of data generation and processing that are utilized in plant breeding have radically changed in recent years. With the advancement of new high throughput technologies, data have grown in terms of quantity as well as complexity. However, the significance of the information that is hidden in newly generated experimental data can often be deciphered only by linking it to other data, collected previously and/or by others. Collaboration that makes it possible to connect disparate data sources and analyze them in meaningful ways with other researchers requires robust but practical data management solutions. CIP has developed an on line Global Trial Data Management System: https://research.cip.cgiar.org/gtdms/. Three key components are CIPCROSS, HIDAP, and Field Book Registry.

CIPCROSS is a botanical seed inventory tracking system for clonal crop breeding that enables tracking of breeding materials from crossing blocks and botanical seed inventories through to seed distribution. CIPCROSS is open source software, available online, and comprises two main tools: (a) The Crossing Management

System (CIPCROSS Tool v1.1 for Pocket PC) used the main development platforms Visual Basic 2008 for Windows Mobile OS and Microsoft SQL Server Compact 3.5 for Windows Mobile software. This tool facilitates the barcode labeling of parents and crosses, storing the information in a database and generating automated reports and (b) The Botanical Seed Inventory System (CIPSIS Tool v1.1) was designed as an easy-to-use web application. It uses PHP, Yii Booster, Bootstrap, CSS, Java Script, HTML and MySQL database programming languages. This tool facilitates the recording and accessing of information on the location of botanical seed in the storage facility, documentation of germination tests, updates on seed stocks, information about crosses and breeder's name. The system helps breeders develop, integrate, and organize their information in a database, avoid typing mistakes and saving work time (24 person–hours are saved per 14,250 labels). The Roots & Tubers Base centralized data repository facilitates access to the information with queries and filters for advanced searches (Fig.—Users can download these tools through the web page of "The Global Trial data Management System from CIP" https://research.cip.cgiar.org/gtdms/)

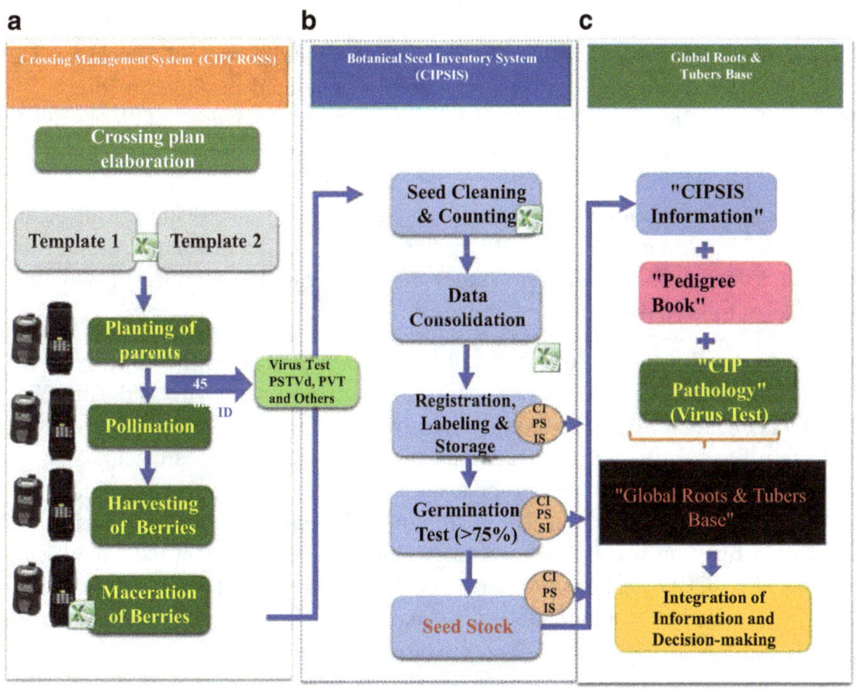

Fig. CIPCROSS workflow: (**a**) The Crossing Management System (CIPCROSS Tool v1.1 for Pocket PC); (**b**) The Botanical Seed Inventory System (CIPSIS Tool v1.1) and c) Global Roots & Tubers Base as a centralize data repository; (**c**) Global Roots & Tubers Base as a centralize data repository

Data collection in potato breeding requires capacity to use and share standard protocols for the selection of clones from seedlings through to observational trials and on to preliminary and advanced yield trials, each accompanied by field books and structures for data collection, analysis, and reporting.

The highly integrated data analysis platform called HIDAP was developed by CIP's breeding program and Research Informatics Unit to facilitate and unify data collection, quality control, and data analysis for clonally propagated crops. HIDAP provides a single platform for use by potato and sweetpotato breeders. It supports compliance with Open Access, open standards such as the potato and sweetpotato crop ontologies and linkages with relevant corporate and community databases such as CIP's BioMart (https://research.cip.cgiar.org/gtdms/biomart) and SweetPotato Base (www.sweetpotatobase.org). HIDAP builds on the statistical platform R. This includes the R shiny tools, the knitr package, the Agricolae package also developed by CIP (https://cran.r-project.org/web/packages/agricolae/agricolae.pdf), and more than 100 other R packages. The R shiny package enables implementation of interactive web pages that are usable online and offline. The knitr package enables the creation of reproducible reports. Numerous statistical analyses can be performed using R and R functions developed at CIP. The software is available for download at https://research.cip.cgiar.org/gtdms/hidap/.

HIDAP is connected to the institutional pedigree and corporate database at CIP facilitating the tracking of clones and families generated through breeding. This connectivity enables verification and maintenance of the identity of clones across the different selection stages and tracing of pedigrees in selection and breeding. The HIDAP network enables researchers to share field books with colleagues, regional breeding programs, and/or partners. To use this network, users must register and create a login account. Once logged in, the user can share, download, and receive field books for different selection stages in a user-friendly interface. A download count helps to keep track of users and uses of this tool.

The Field Book Registry (https://research.cip.cgiar.org/cipfieldbookregistry/) facilitates updating field books in real time and viewing their status in the database. The data generated in CIP's potato breeding program is stored in the *"Global Roots & Tubers Base*" utilizing the free BioMart software https://research.cip.cgiar.org/gtdms/biomart/. This database has been structured for storage of phenotypic, genotypic, pedigree, geographical, and environmental data. Through the metadata and the search function using filters, the user can retrieve data from the experiments conducted by CIP or partner programs using CIP materials. The availability of the data is managed in conjunction with Dataverse following CGIAR open access guidelines.

The use of the database and software tools enables analysis of phenotypic and genotypic data. A key goal is the identification of effective models that predict phenotypic traits and outcomes, elucidating important biomarkers and generating important insights into the genetic underpinnings of heritability.

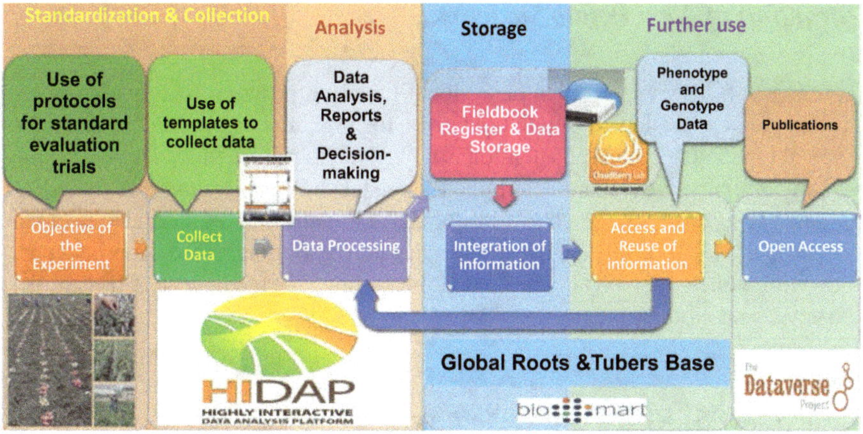

Fig. Flowchart of CIP's breeding information system: Standardized evaluation protocols and ontologies implement templates to collect and analyze data with HIDAP software; Subsequent data storage through the field book register allows incorporation of breeding information into the Roots & Tubers Base in which phenotypic and genotypic data are integrated and made available for further use

Setting Objectives for Potato Breeding

Successful breeding relies fundamentally on having a clear set of prioritized breeding objectives. Key enabling information and tools for setting breeding objectives include: knowledge of the opinions and needs of stakeholder groups, knowledge of the production constraints and cropping systems of target populations and environments, and standardized means to measure and describe traits. Knowledge of the strengths and weaknesses of predominant varieties provides an important baseline for setting breeding objectives.

Having a good understanding of desired end-user traits requires ongoing, interdisciplinary communication and collaboration. This helps increase the chances that a new variety meets actual needs and is thus adopted, as well as minimizing the chance that new varieties (and the traits that define them, their management and marketing or use) might marginalize or disadvantage social or gender groups such as poor farmers or women, who often have access to different resources than men do. There is also a need to consider social and economic dimensions and client demography such as market access, youth, and urban/rural factors. A CGIAR system-wide initiative is currently underway to assess and encourage breeding that responds to social diversity including often-neglected or marginalized groups with emphasis on gender (http://www.rtb.cgiar.org/gender-breeding-initiative/?lang=en).

Potato breeding is characterized by having a large number of objectives, which include traits specific to the intended market and use of the crop (end-user traits), as well as traits related to productivity and protection against local diseases and insect pests (grower traits).

Targeting of New Potato Varieties

Building on CIP's experiences of a global breeding program that developed two agro-ecologically adapted populations (one for highland tropics and one for subtropical lowlands), increased attention is now given to prioritizing packages of traits for which relative values are determined with stakeholders in a more decentralized manner. This involves the cross-disciplinary estimation of trait values at global, regional, and local scales, the development of corresponding variety or product profiles with stakeholder involvement, and the setting of breeding priorities to meet those via selection decisions.

Case Study 5: RTB Priority Setting

An example of assessing trait values can be seen in the priority setting exercise conducted by the Root, Tubers and Bananas (RTB) Program of the CGIAR (http://www.rtb.cgiar.org/resources/impact-assessments/) Kleinwechter et al. (2014). Through a global survey carried out from 2012 to 2013, the RTB program sought to rank production constraints of each RTB crop (i.e., cassava, yam, banana, sweetpotato, and potato) and research options to alleviate them. This exercise was led by agricultural economists specialized in impact assessment. It provided an overview of problems affecting the potato sector and potential solutions including, but not limited to, breeding. The potato survey (Hareau et al. 2014) provided respondents with a list of 91 research options, organized around the areas of crop improvement, crop and resource management, seed management, genetic resources, value chains, postharvest utilization, and marketing, as well as socioeconomic research and extension.

Five of the top ten ranked research options for potato were directly related to breeding. Breeding for late blight resistance, drought tolerance, earliness, and high yield ranked second, third, fourth, and eighth of the 91 options provided, while the generic "germplasm enhancement" ranked seventh. An ex-ante assessment of the selected potato technologies revealed significant differences in terms of net present value (NPV) and internal rate of return (IRR) on investments across the different research options. Late blight- and virus-resistant varieties had the largest expected net benefits ($US 4.7 billion and $US 3.9 billion), and high rates of return of 87% and 104%, respectively, in the high-adoption scenario.

Case Study 6: Adjusting and Ranking Priority Traits in New York (NY) State

Staff of the breeding program in New York State U. S. meet many times each year with stakeholders, in both formal and informal settings, to continually discuss what the program's priorities should be. The frequent interaction helps stakeholders feel comfortable expressing their views, and allows the program to detect changes in industry priorities more quickly, as breeding goals always change with time.

Two examples can illustrate how stakeholder feedback has changed the NY program over the past 5 years. (1) NY growers recently made repeated requests to develop earlier maturing potatoes, to solve a problem the NY breeding program had unwittingly helped create. NY breeders have always put a high priority on yield, and selection for yield tends to select for later maturity. Unfortunately, a suite of late maturing varieties also shortened the timeframe in which NY growers could harvest their crop. The NY program now pays much more attention to maturity, and is prepared to select potatoes with lower yield if they mature early. (2) Potato chip factories recently began to ask for smaller chipping potatoes, as more and more of their product is now sold in small bags. In response, the program now selects smaller potatoes than it (or regional chipping factories) would have been willing to process in the not-so-distant past.

One approach that the NY breeding program has found useful when prioritizing traits is to compare what stakeholders ask for ("what stakeholders say") with the attributes of widely grown varieties ("what stakeholders do"). When there is an apparent disparity between words and action, there is an opportunity for deeper understanding. Fifteen years ago growers in NY kept asking for new varieties with resistance to common scab. What made the request unusual is that growers already had several resistant varieties to choose from, and that the variety they grew most each year was highly susceptible. What the NY program eventually realized is that the popular scab susceptible variety had two quality attributes—outstanding fry color out of cold storage and high specific gravity—that were far more important to the chip factories than the scab resistance growers kept asking for. As a result, NY re-ranked selection criteria, placing fry color and specific gravity above resistance to common scab.

Product Profiles

A product profile establishes a set of targeted attributes that a new plant variety or animal breed is expected to meet for successful release in a given market segment. Attributes must be understood as traits reaching a specified level; this level being defined either in absolute or relative terms (Ragot et al. 2018). Thus, a product profile may list yield (25 tons/ha or more; or 15% over variety V1 across a range of soil fertility conditions), tolerance to potato leaf roll virus (same as or better than variety V2), or dry matter content (no less than 18%, or no less than variety V3). The development of product profiles may best be done in collaboration with, for example, pathologists, agronomists, or nutritionists, as well as user communities who can contribute specific knowledge, tools, and approaches for setting trait levels, which in turn become the breeding targets within a profile.

Tools and Metrics

CIP's potato breeding program targets low input conditions by relying heavily on endemic disease pressure, poor soils and a series of intentional exposures to stress for screening and selection. The product profiles incorporate quantitative breeding objectives for productivity, protection, and utilization traits with emphasis on resilience, the setting of quantitative breeding objectives relies on knowledge of baseline variety characteristics and available genetic resources, the expected effects of changes in trait levels, and means to measure gains toward them.

Metrics for disease resistance: One special tool used by CIP is an interval scale for expressing potato resistance to late blight that indicates resistance levels required for satisfactory control of the disease in agro-ecologies with varying degrees of pathogen pressure. The use of this interval scale facilitates setting quantitative resistance breeding objectives in a robust manner within a breeding program's trait improvement framework. The late blight susceptibility scale of Yuen and Forbes (2009) enables assessment of resistance levels with a reduced coefficient of variation among trials as compared to other semi-quantitative metrics like AUDPC or rAUDPC. The scale uses reference cultivars in regression analysis and helps breeders to measure and describe resistance of genotypes independent of environment or inoculum level, which can vary from site to site and year to year. This approach gives breeders a simple numerical metric for quantitative traits that is useful for setting baselines, breeding goals, and calculating genetic gains.

Application of the susceptibility scale to setting breeding targets is based on the understanding that a variety with level 5 (more susceptible) would be sufficiently tolerant to provide the same level of protection in low-pressure agro-ecologies as a variety with level 2 (less susceptible) would provide in a agro-ecology with high disease pressure. The use of this scale is illustrated in Field assessment of resistance in potato to *Phytophthora infestans* at https://research.cip.cgiar.org/potatoknowledge/lateblight.ph.

Metrics for nutritional traits: To help reduce the health burden of iron and zinc deficiencies, quantitative targets for levels of these two elements in potato (a biofortification breeding goal) were set through collaboration between potato breeders at CIP and nutritionists of Harvest Plus. Consideration was given to (1) nutritional status of the target population, (2) dietary increments of iron and zinc known to be effective from other approaches (i.e., food fortification or nutrient supplementation), and (3) features affecting feasibility of nutrition impact such as consumption (potato intake), heritability, bioavailability, and retention of these minerals in potato prepared for use as food.

Nutritionists consider that dietary increments of 0.4 and 0.2 mg/day of iron and zinc, respectively, can have a positive biological effect on the health status of populations at risk of micronutrient malnutrition associated with deficiencies in these minerals. The provision of these increments through biofortified varieties is feasible when mineral concentration of a variety, consumption levels, bioavailability, and retention in the diet are sufficient. This food-based approach to reducing micronutrient malnutrition relies on variety change but not on modification of consumption patterns, i.e., high versus low potato intake. Logically, a population that consumes twice as much potato as another one would ingest twice as much iron from that source. And the percent of the Estimated Average Requirement (EAR) of the nutrients it provides would be double.

Quantification of these features for a given varietal and dietary context assumes conservative estimates of 10% and 25% bioavailability of iron and zinc, respectively, from potato, and that minerals are not lost in cooking. Figure illustrates how different levels of iron and zinc concentrations in potato (referred to as biofortification levels: Base line, 1, 2, and 3, where base line represents current potato varieties) contribute to the EAR of iron and zinc for women of fertile age who consume 100, 200, or 400 g of potato per day. For reference, women in parts of Rwanda consume over 500 g of potato per day (Personal communication, Harvest Plus), while women in parts of Peru, may consume 800. The iron and zinc levels included in potato product profiles for populations of the Andes or the central African highlands are 45 mg/kg Fe and 35 mg/kg Zn (just over biofortification level 2 in Fig) which can be expected to provide 50% of the EAR of both minerals for women consuming 400 g potato a day. This is a considerable increment over the baseline, but has been assessed as feasible by CIP's breeding program after evaluating genetic diversity, estimating heritability, and realizing significant gains (subsection 6.1.5) in cultivated potato germplasm. Achievement of such genetic gains would only provide 12–15% of the EAR for women who consume 100 g of potato per day or less, as in much of the subtropical lowlands of Asia. The inclusion of a biofortification breeding target to increase Fe or Zn intake from potato by 0.4 or 0.3 mg/day, or reach 50% EAR for this market segment, would require greater genetic gains than presently estimated to be feasible by the interploid breeding strategy undertaken by CIP. Nevertheless, high iron and zinc potatoes have been requested by the national programs of both India and Bangladesh. In such cases, a

A

Biofortification Level	Iron (mg/Kg DW)	Zinc (mg/Kg DW)
Baseline (non-biofortified)	20	16
1	27	24
2	42	32
3	50	40

B

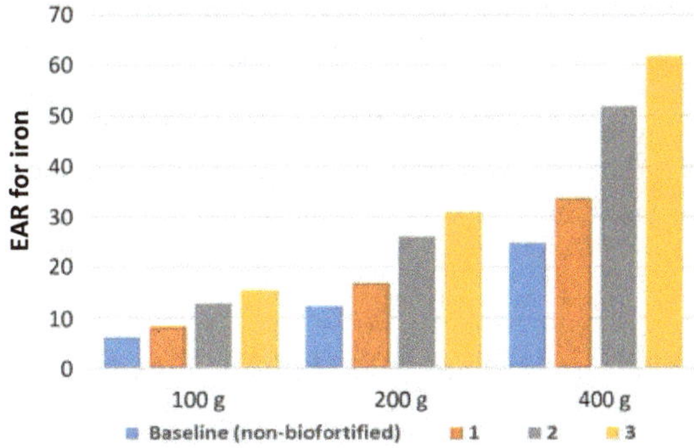

Fig. Influence of Fe and Zn concentration and intake of potato on the estimated average requirement for women of fertile age. (**a**) Definition of baseline (non-biofortified) and incremental concentrations (biofortification levels 1, 2, and 3) of iron and zinc in potato tubers. (**b**) Percent of EAR for iron for women of fertile age met by consuming 100, 200, or 400 g/day of potatoes with concentration levels 0, 1, 2, and 3

food systems intervention seeking overall increments of micronutrients from co-staple crops should be considered.

Biofortification targets between level 2 and 3 in Fig. have been set as part of the product profiles for table potatoes oriented to tropical highland agro-ecologies and populations with significant levels of anemia, who also consume potato as a main food or staple crop. The value of nutritional traits can be assessed with the disability adjusted life years (DALY metric). DALY is ordinarily used to assess the impact of public health burdens like human disease or illness. It extends the concept of potential years of life lost due to premature death to include equivalent years of "healthy" life lost by virtue of being in a state of poor health or disability. In so doing, mortality and morbidity are combined into a single, common metric (Meenakshi et al. 2007).

Setting of Breeding Priorities

Once product profiles are defined, breeding programs can proceed to refine trait lists and allocate resources to address those that are most feasible. While goals and priorities are important for breeding programs and the institutions that host and sponsor them, the list of traits a breeder can work with is limited, and in this light, he/she will select a subset of those in the profile for the day-to-day efforts of the program. Breeders may also be aware of a few "must have" traits that do not come up in stakeholder consultation, and since the inclusion of any trait in a breeding effort (whether to change or even maintain trait levels) requires investment, and thus has a cost, these should be made explicit and, ideally, added to the product profile.

Experience and knowledge of existing germplasm are applied to identify the set of attributes to be addressed in breeding. For each attribute, a quantified description of the desired result (maximize, reach specified level, maintain a certain minimum level, etc.) and a unique rank or priority is assigned. Ranking represents the relative effort required to develop the set of traits that will enable the new variety to respond to the opportunity or constraint identified in the targeting exercise and taken on by the breeding program. Relative effort required to address each objective are determined by the proximity of the program's genetic materials to the needed level of each trait, and the genetic control and relationships among various traits.

Selection Decisions in Potato Breeding

Because potato is highly heterozygous, a large number of traits segregate every time a cross is made. Every offspring receives a unique combination of desirable alleles at some loci and undesirable alleles at others. No offspring is ever "perfect." An issue of considerable practical importance, therefore, is deciding which phenotypes (clones; genotypes) are worth keeping, and which should be discarded at each stage of the selection scheme.

Selection decisions are the decisions a breeder makes to use, advance, or discard a selection unit (plant, progeny, progenitor, clone, experimental variety), guided by the breeding priorities, using specific decision-making tools (e.g., assessments of the targeted traits, estimates of breeding value, presence/absence of specific markers, selection indices, combining ability).

In early stages of selection, when the number of plants of each genotype is small, it is important to select stringently only for highly heritable traits. Meaningful assessments of yield, on the other hand, can only be performed in later stages. As selection progresses, and the individuals not meeting breeders' criteria have been eliminated, it becomes especially important to keep product profiles in mind, i.e., to select not just on the basis of individual traits, but on the suite of traits that collectively define a product, with attention to specific traits at specific stages of the selection scheme.

Case Study 7: Profile and Selection Decisions for Chipping Potato for Northeast USA

As nations develop there tends to be a shift in potato consumption, away from fresh and more towards processed potatoes, primarily as chips or French fries. Varieties intended for fries or chips ideally have higher levels of starch, so that they absorb less oil when fried. In addition, if potatoes need to be stored in the cold for a long time—typically to ensure a year-round supply of raw product—then it is useful if they are resistant to cold-induced sweetening, as the presence of high levels of glucose and fructose will cause potato slices to turn dark brown when fried.

To improve resistance to cold-sweetening, the New York program has adopted a low-tech approach that could be adopted anywhere. The essence of the approach is this: store potatoes at a temperature where only a small percent will fry to a light color. Intercross those that do, and after 5 years or so, once the population has reached a point where many progeny fry well, lower the cold storage temperature by 0.5–1.0 °C the following year. Intercross the few that fry well under the new, colder regime, and repeat the process. Recurrent selection has proven very effective at improving fry color in the New York program; many clones now fry well out of 3–9 months of storage at 6 °C. Ideally the breeding program will fry from slightly colder storages than those currently in use by regional industry, to help drive down the temperatures that industry can use over time. In general, the colder the storage, the longer that potatoes can be stored.

The two most important attributes in chipping potatoes are starch content (strongly correlated with specific gravity) and fry color, followed by tuber shape and size. About 70% of the New York breeding program effort is dedicated to developing new chipping varieties.

The New York program practices very little selection on seedling tubers, as the performance of plants grown in pots correlates poorly with performance of plants grown in the field. When the first field generation is harvested, the principal selection criteria are for tuber yield, tuber shape, and size (all visually assessed). The ideal chipping potato is round and the size of a baseball. Clones with low yield, or shapes and sizes too far from baseballs, are not selected.

The New York breeding program chips (out of cold storage) every clone that survives 1 or more years of selection in the field, discarding all that do not chip as well as, or better than, the industry standard chipping variety "Snowden." Absolute fry color can vary from season to season, although the ranking of fry color between clones does not vary greatly.

In the USA a specific gravity of 1.080 is deemed the minimum for processing. Nevertheless, because starch content can vary considerably from season to season (warm nights result in lower starch content than cool nights, all other things being equal), in New York the level of starch is always evaluated relative to that in the widely grown chipping variety, "Atlantic." The specific gravity of Atlantic is typically above 1.090 in New York, but in some seasons it is less. The New York breeding program will not continue to evaluate a chipping clone whose gravity averages 0.010 or more less than Atlantic, and prefers clones to average 0.005 or fewer points less. New York begins to evaluate specific gravity after a clone's second year in the field.

Replicated, multisite yield trials begin in the fourth field year, where clones that yield less than 90% of Atlantic are discarded. Clones that exhibit considerably more internal or external defects than "Atlantic" or "Snowden" in yield trials are also discarded, regardless of their yield, fry color, or specific gravity. Clones that survive several years of field trials are then evaluated on commercial farms, and processed in commercial chipping plants, where growers and chipping plants make the final decisions about which clones merit release as new varieties.

Selection Decisions: Marker-Assisted Selection

How the New York Program Uses Molecular Markers

Even though current NY breeder WDJ was trained as a molecular geneticist, the New York breeding program does not make extensive use of marker assisted selection, primarily because there are not yet many publicly available markers linked to traits of high priority.

One of the two markers currently used in New York is 57R (Finkers-Tomczak et al. 2011), tightly linked to the H1 gene, which confers resistance to race Ro1 of *Globodera rostochiensis*. Developing potatoes resistant to race Ro1 has been the highest priority of the New York breeding program for the past four decades; almost every cross made in NY has at least one Ro1-resistant parent. All offspring that survive 2 years of visual selection for appearance and yield in the field (about 200 clones, out of 20,000 initially planted) are screened with 57R over the winter. The vast majority of clones lacking the marker are discarded, the only exceptions being those that look especially promising for other high value traits (e.g., excellent chip color combined with high specific gravity).

RYSC3 (Kasai et al. 2000) is the other marker currently used in New York. This marker is tightly linked to Ry_{adg}, which confers immunity to potato virus Y (PVY), and like 57R, is used to screen clones that have survived 2 years of selection in the field. At present the presence/absence of the RYSC3 marker is just one of many data points NY uses to make a decision about each clone, with other data points including chip color, specific gravity, resistance to common scab, and visual assessments of yield and tuber appearance.

Although markers tightly linked to potato virus X (Gebhardt et al. 2006), wart (Gebhardt et al. 2006), verticillium wilt (Bae et al. 2008), Columbia root-knot nematode (Zhang et al. 2007), the pale cyst nematode (Sattarzadeh et al. 2006), and late blight resistance genes (e.g., Colton et al. 2006) have been reported, none of these traits are important enough in New York to influence selection decisions, and thus none of these markers are used routinely.

Breeding Objectives at CIP

Priority setting exercises (Fuglie 2007; Hareau et al. 2014), complemented by evolving requests from national programs for elite germplasm and candidate varieties, both contributed to CIP's profiles for new potato varieties tailored to the contrasting agroecologies, cropping, and food systems its program addresses. Two key products conceived in a recent exercise were (1) "Agile Potato" for intensification and diversification of cereal-based systems of Asia and, (2) mid-elevation tropical highlands late blight resistant potato.

CIP's major breeding populations are oriented to develop table and processing varieties for sustainable potato production in tropical highland and lowland areas faced with either of two basic sets of biotic and abiotic constraints: those encountered in the tropical and subtropical highlands, where late blight, cyst nematodes and frost are limiting, and farmers are increasingly faced with drought; and warmer, drier production areas, that are challenged by viruses, bacterial wilt, drought, root knot nematode and other pests, Increasing requests to support production in mid-elevation tropics of Sub Saharan Africa and Asia as well as temperate regions of Asia have led to the inclusion of heat tolerance and long day adaptation into our breeding objectives, requiring the introduction of new germplasm, screening methods, and trial sites. Present emphasis is on: earliness to enable potato's full potential to contribute to incomes and productive cropping systems; heat and drought tolerance to enable expansion to mid-elevation regions nearer to population centers, at the same time helping farmers adapt to climate change, and; nutritional value with specific attention to iron and zinc biofortification. CIP aims to reach new thresholds for these traits, while maintaining and broadening levels of disease resistance achieved in the past, and also responding to regional preferences for culinary characteristics and market opportunities.

Case Study 8: Agile Potato for Asia

Product Profile

The "Agile Potato" project seeks to develop potato varieties for fresh and processing use in short crop production windows of Asia. The product profile includes 70–90 day maturity; resistance to multiple viruses (PVY and PLRV, with PVX resistance desirable); stress tolerance (heat, drought, and salinity); resistance to late blight; and market-oriented quality (high dry matter and chip processing). The expectation is that varieties with short crop duration that resist disease, require less-frequent seed replacement, thrive in warm seasons, and store well under rustic conditions, can readily be inserted into diverse cropping patterns dominated by rice or wheat. Agile potato varieties would fit into cropping windows currently left fallow in the cereal-based systems of subtropical lowland and temperate regions of Asia. They should help increase system productivity through intensification as well as help meet growing demand for processed potatoes, thus contributing to income generation and employment. Breeding to meet this profile is backed up by trait research to enhance and accelerate breeding methods, and inter-disciplinary research to promote fast-track systems for variety identification and release. With this profile, CIP seeks to develop strategies for sustainable intensification of farming systems, raise awareness and expand consumer demand for potato while diversifying cereal-based diets with potato as a nutritious co-staple or vegetable food. A gamut of research on ecological and socioeconomic consequences of intensification is also planned in order to identify appropriate indicators of sustainability and monitor effects of benefits to intended sectors, including the poor whose livelihoods would be affected by the proposed intervention.

The target is to select 70–80 day heat-tolerant potatoes with competitive yields and more than 18% dry matter and resistance to potato viruses for subtropical lowlands of Southwest Asia. Commercial varieties and elite selections from previous cycles of the lowland tropics virus resistant population (LTVR) are comparators for performance in representative or target selection sites. In the following figures we show a group of selected early clones that performed significantly better than past generations and local varieties for yield and quality at 70 days after planting in warm environments, yielding in a range of 20–40 tons/ha, compared to the best local varieties such as Unica and Reiche and adequate dry matter content over 18% under the abiotic stress conditions of lowlands.

Dry matter (DM) content is an important component to be taken into account in the breeding process especially when selection is focused on lowland tropical areas. In addition to the loss of fresh weight yield of tubers associated with high temperatures, the DM content of tubers are also reduced by 1% as the air temperature rises by 1 °C over a temperature variation average of 15–25 °C. This is due to the rapid

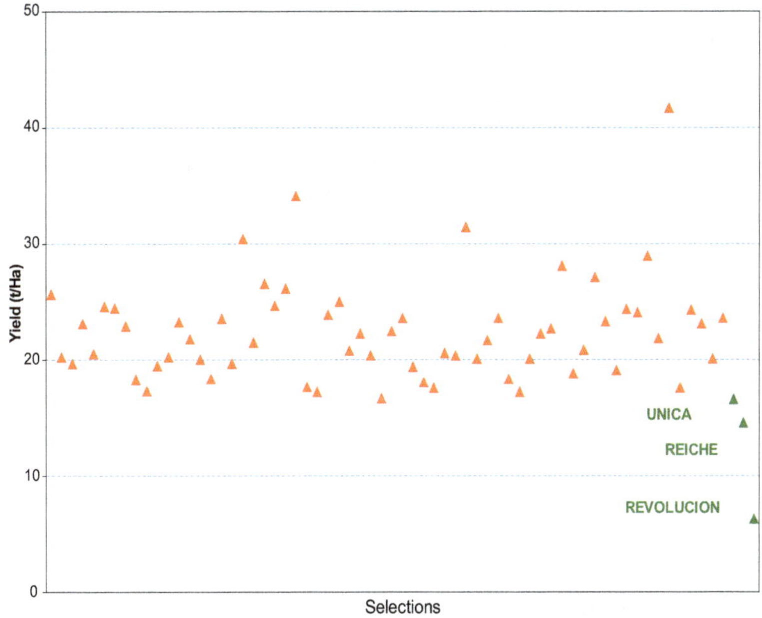

Fig. Scatter plot showing yield of selected clones in a new generation of LTVR population that surpass levels of varieties and clones from previous generation grown under hot conditions of San Ramon, Junin, Peru, and harvested at 70 days after planting

decline of photosynthesis (Midmore and Rhoades 1988). This tendency of decrease in DM of clones in the LTVR population can be appreciated in Fig., showing decreasing DM contents with decreasing altitude and thus increasing temperature of the evaluation sites. However, breeding by recurrent selection for yield as well as DM content in this population has maintained this important trait at between 11 and 19% even in the warmest season, and 15–21% in the high humid jungle site of San Ramon, Peru. This means that genetic diversity is sufficient to permit selection for heat tolerance and genotypes with good yield and adequate DM content can be identified in LTVR in spite of the critical negative correlation between temperature and DM in potato.

Crossing with long day adapted parents and evaluation under simulated long day conditions in Peru permitted the identification of LTVR families and advanced clones with adaptation to higher latitudes, for example of Vietnam where day lengths are intermediate, and Uzbekistan and Tajikistan where they are long combined with high summer temperatures. The same screening approach also helped provide pressure for improving the earliness (70–80 days) of LTVR at low latitudes. The compensatory effect of temperature on day length provided by warm mid-latitude conditions of southern Peru has also been key in breeding for early maturity, which can also be referred to as a relaxed requirement for short days. The identification of a moderate frequency of genotypes as well as superior parents with good performance under higher latitudes suggests that the LTVR breeding strategy

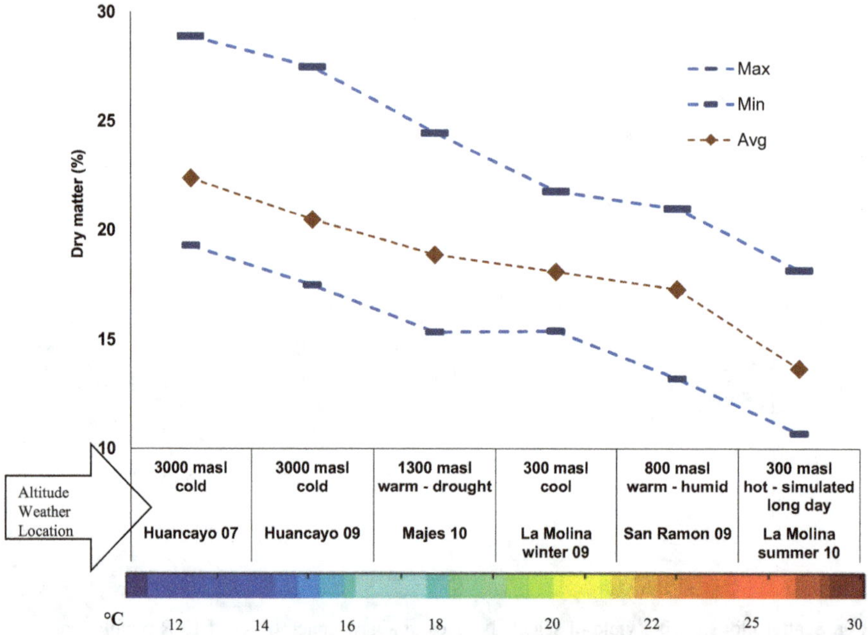

Fig. Variation for dry matter content of potato selections in LTVR population across environments. Three hundred LTVR clones were grown in ten hill plots in six locations in Peru and evaluated for dry matter content within 10 days of harvest

is likely to be successful in developing the agile potato for both of its target environments—subtropical lowland and temperate low and mid-elevation regions of Asia.

Source Population LTVR

From the 1980s through the late 1990s, CIP endeavored to strengthen its lowland tropics (LT) population with multiple virus resistance developing the lowland tropics virus resistant population (LTVR). Development of the broad-based LTVR population used an open-recurrent selection strategy to incorporate foreign elite and advanced clones into its lowland tropics adapted breeding population. Lowland tropical environments were characterized as having day lengths of 10–14 h, minimum night-time temperatures of 18–20 °C, mean maximum temperatures greater than 25 °C and mean annual soil temperatures at 50 cm depth of 22 °C or more. LTVR population combines the heat tolerance and early bulking ability of *S. tuberosum* germplasm bred under the summer conditions of the northern Hemisphere (Cubillos and Plaisted 1976) on one hand, with virus resistance from native *Andigenum Group* germplasm and long day adapted *Andigenum Group* germplasm, namely Neotuberosum, on the other (Plaisted 1987; Mendoza 1990). The Neotuberosum and *Andigenum Group* germplasm contributed major genes for extreme resistance to potato virus Y (PVY) and potato virus X (PVX) (Muñoz et al. 1975; Gálvez et al. 1992). Early generation screening was routinely performed to discard PVX and PVY susceptible individuals in any trial in which LTVR families were evaluated.

Testing sites included CIP's experimental stations in Peru which are relatively dry, humid, hot, and hot and humid lowland environments. Experiments comparing the yield stability of diverse clones across these locations revealed that hybrids from *S. tuberosum* × Neotuberosum crosses performed significantly better than those from *S. tuberosum* × *S. tuberosum* or *S. tuberosum* × *Andigenum Group*. It is possible that the higher level of heterozygosity together with prior selection of the divergent parental materials for adaptation accounted for the better performance of crosses between *S. tuberosum* and Neotuberosum clones (Mendoza and Estrada 1979). These authors also reported that *S. tuberosum* × neotuberosum hybrids are adapted to hot and humid conditions and able to produce tubers at 60–70 days after planting, albeit under short days. These hybrids made a strong contribution to the LTVR population, and thus may account in part for its adaptation to warm conditions. The subsequent cycles of genotypic recurrent selection led to the selection of a number of multiplex PVY and/or PVX extreme resistant progenitors (Mendoza et al. 1996) some of which have since been shown to also possess good GCA for yield and dry matter content. Multiplex refers to the genetic constitution of a polyploid individual having more than one copy of an allele, in this case RRrr or RRRr, assuring the transmission of resistance conferred by a major gene to nearly 100% of its cross progeny. In parallel to population improvement and parental line development, high yielding clones with excellent tuber quality were selected during each recurrent selection cycle.

A second stage of the breeding process aimed to incorporate foreign elite and advanced clones by means of an open-recurrent selection strategy. Assessment of genetic parameters in a North Carolina Design II estimated heritability for resistance to PLRV infection to be 0.54–0.69. Higher estimates were obtained when greenhouse-reared infective aphids were used to conduct controlled inoculations with sprouted tubers before tuber families were planted in the field, suggesting that controlled inoculations minimize errors or escapes because of a more even distribution of inoculum. Additive genetic effects were shown to contribute the most to resistance variability (74%), followed by dominance effects (10%) and additive × environment (location) effects (5%), suggesting that breeding advances could be expected in further cycles of recurrent selection (Salas 2002).

Multilocation Testing and Selection Decisions

A stage plan can be used to communicate selection decisions made in early and late stages of breeding. The stage-gate process portrays stepwise decisions with thresholds for each characteristic that govern decisions to keep or discard individuals in the course of a given breeding cycle. In this process, the "gates" describe the collection of thresholds or criteria that must be met if a clone is to pass from one stage of selection to another along the clonal selection scheme outlined in Fig. , presents a stage plan for selection in the LTVR population to meet the Agile Potato profile.

Yield and processing quality traits of LTVR clones generated during open-recurrent selection were evaluated in replicated experiments in a range of lowland environments of Peru. Dry matter content was consistently higher in the cool arid lowlands of the coast under irrigation (average = 21%) than in the warm humid area of the central jungle (average = 19%), highlighting the detrimental effect of high temperatures on this trait. As expected, significant G × E interaction was encountered for dry matter content, but also for glucose content and chip color. High yielding clones with good table or processing quality combining virus resistances and heat tolerance have been released by national programs and public institutions in target countries. Varieties of LTVR origin include: Tacna, Costanera, Unica, Reiche, Maria Bonita, and Maria Tambeña released in Peru; Kinga, Meva, Kinigi, Muziranzara, Muruta, Yayla Kizi, Baseko, Enfula, Chamak, Dheera, and IRA-92 in Africa; and Raniag, which was named in the Philippines. Evidence of the wide adaptation of bred materials from this population is provided by the success of the variety "Tacna" in the lowland subtropics of China where it is known as "Jizhangshu 8" and planted on an estimated 133,215 ha in 2015 from which 70,533 ha in Hebei, 26,000 ha in Shanxi, 22,667 in Inner Mongolia, and further 3000 has in the increasingly drought-prone northwest province of Gansu with more than 3000 ha (Gatto et al. 2018). Additional evidence of broad adaptation comes from the variety Unica that is now grown in at least six countries.

Breeding Priorities

Following CIP's multidisciplinary conceptualization of the agile potato, the potato breeding program refreshed its strategy by setting priorities for combining the needed traits in subgroups of the LTVR population. The inclusion of resistance to PVY in the LTRV population and selection for highland tropical as well as subtropical lowland conditions to enable seed production has had a cost in terms of earliness even after three to four cycles of recurrent selection. The infusion of Tuberosum germplasm and selection under heat and mid-long days, for plant type and stability from early generations is strategy to improve earliness in the LD (long day) and HT (heat tolerant) groups. In this interest, tuber quality becomes an issue and attention is needed to keeping dry matter high and glycoalkaloids, which provide a bitter flavor to potato and can be toxic, low.

Often unrecognized and rarely controlled, the accumulation of viruses in seed and ware tubers is perhaps the most serious constraint to economical potato production in the tropics. Breeding for resistance to the world's most important potato virus PVY is possible due to the availability of major genes (Ry_{adg}, Ry_{sto}) that confer extreme resistance, and for which selectable markers are available. On the other hand, developing effective levels of resistance to PLRV is one of the more challenging components of the Agile Potato profile. Most sources of resistance to PLRV are quantitatively inherited and provide partial resistance to infection and/or accumula-

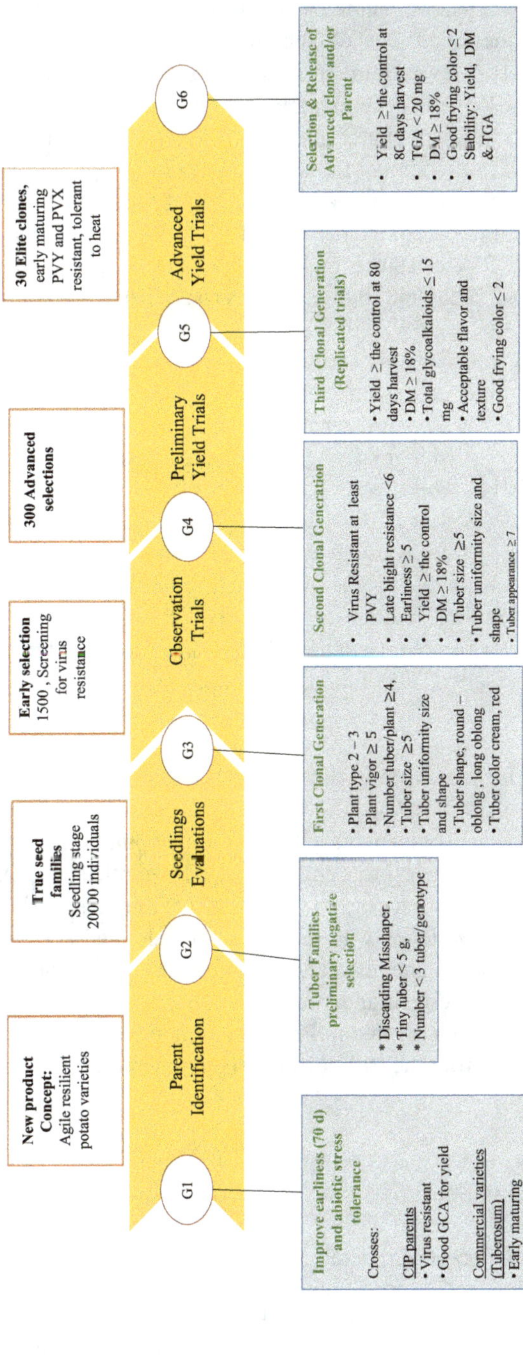

Fig. Stage plan for selection in the LTVR population to meet the Agile Potato profile

tion of the virus. Despite moderate heritability, the frequency of resistant clones in advanced populations has been low. Thus, a relatively high priority is given to incorporating a new, highly heritable source of high levels of resistance to PLRV, designated Rl_{adg} recently identified in Andigenum Group germplasm into advanced populations that already carry extreme resistance to PVY at high frequency.

New sources of PLRV resistance, namely, accessions LOP-868, HUA-332 and OCH-7643, confirmed following mass selection in *S. tuberosum* Andigenum Group germplasm held at CIP, were shown to have exceptionally high GCA for this trait (Mihovilovich et al. 2007). Further characterization of these sources has been the basis of supporting trait research, the benefits of which are expected to boost levels and frequencies of PLRV resistance in the LTVR and other potato populations. Using a population of dihaploid individuals derived from LOP-868, Rl_{adg} was mapped to potato chromosome V, with high heritability demonstrated to be due to its multiple resistance mechanisms and a duplex allelic conformation of a major dominant gene in the donor accession (Velásquez et al. 2007). A PCR based marker (RGASC850) located 2 cM from the gene has been developed to facilitate the introgression and tracking of this major gene in advanced breeding populations (Mihovilovich et al. 2014). This marker has been incorporated into a multiplex PCR system for assisted selection of PLRV and PVY resistance using previously available information on Ry_{adg} from Kasai et al. (2000).

Table presents the attributes, profile and breeding priorities contributing to the Agile Potato program and outlines some of the key research that supports the objective. Breeding priorities represent the relative effort dedicated to each attribute of the profile, given present status of the LTVR population.

Multi Trait Selection for Increased Earliness and Tolerance to Abiotic Stress

Most recently, emphasis in improving CIP's LTVR population toward the agile potato has targeted increased earliness and tolerance to abiotic stress. A new earlier maturing and abiotic stress tolerant generation that combined superior LTVR progenitors with tuberosum varieties was generated. Accelerated selection based on multilocation trials and simultaneous screening for abiotic stress (heat, drought and simulated long days) in divergent and stressful environments as well as screening for PVY resistance was implemented in Peru. The scheme requires only 4 years of selection and evaluation from the true seed progeny generation until the identification of advanced clones.

Evaluation and Selection Sites

CIPs proximity to significant agroecological diversity in Peru is an asset for the identification of patterns and trends of potato genotypes' response to environments. Potato is grown in coastal, high jungle and highland ecologies ranging from sea level to 3500 masl, providing wide variation in weather patterns including tempera-

Table Agile potato profile, breeding objectives, and supporting research

Attribute	Product profile	Breeding priority
1. Yield 2. Earliness 3. Table quality 4. PVY resistance 5. PLRV resistance 6. Heat tolerance 7. Good storability 8. Late blight resistance 9. PVX resistance 10. Water productivity 11. Cold chipping ability 12. Bacterial wilt resistance	1. >25 tons/ha 2. 70 day maturity 3. TBD (flavor appealing to consumers, more than 17% dry matter, <20 ppm glycoalkaloid content) 4. Extreme resistance to PVY 5. Tuberization at >22 °C night temperature; bulking under warm day temperatures 6. Drought tolerance-TBD (ratio of fresh tuber yield to applied water expressed as kg/ha/mm) 7. Bacterial wilt resistance-TBD % plants wilted to degree *x* and % latent infection 8. Resistance to PLRV as high or higher than Granola 9. TBD (*x* days to sprouting, low rottage, low water loss at given temperature) 10. Late blight resistance <6 on susceptibility scale 11. 20% selections with chip color score 1–2 after cold storage; dry matter >18% 12. Extreme resistance to PVX	1. Earliness 2. High and stable yield across day lengths 3. PLRV resistance 4. Heat tolerance 5. Flavor and dry matter 6. PVY resistance 7. Water productivity 8. Storage quality 9. Chip quality 10. PVX resistance 11. Bacterial wilt resistance

Supporting research:
Earliness: implement protocol and validate markers to develop MAS for bulking-based maturity. Complete GWAAS for day neutrality in subset of Trait Observation network panel across environments
Introgress and increase frequency of Rl_{adg} using MAS for PLRV resistance
Implement proximal sensing phenotyping for drought tolerance (water productivity). Develop inexpensive assay for glycoalkaloid contents
Validate pre-bred *S. commersonii* source of bacterial wilt resistance in Kenya and Ethiopia
Refine genome estimated breeding values (GEBV) for early bulking and heat tolerance extend pilot study of genomic selection for early bulking to LTVR in Asia
Optimize field trials and statistical analysis to make full use of trial data; genetic gains and heterotic groups at tetraploid and diploid levels

ture, humidity and rainfall. Most environments have short daylengths of around 11–12 h but the southern extreme at 17°S, provide daylength of about 14 h, which considering high summer temperatures approach threshold conditions for long day adaptation. Three experiment stations (San Ramon, Huancayo and La Molina) and additional sites accessed by local agreements (Table provide evaluation conditions representative of the tropical lowland and highland potato environments targeted by agile potato. Indeed, potato seed for the subtropical lowlands is often produced at higher, vector-free, locations, requiring adaptation of the same variety to both highland and lowland conditions. Moving potato populations through the biotic and abiotic stresses present among these sites, and particularly the combination of temperature and day length, permit the identification of potato genotypes with wide adaptability and stability for yield.

Table Location and conditions of major selection sites used for potato breeding by CIP in Peru

Location	Season	Climate	Latitude (°S)	Altitude (masl)	Geography	Main traits assessed	Agroecology
San Ramon (SR), Junín	Winter	Warm, humid	11	800	Central jungle	Heat tolerance; glycoalkaloid accumulation	Subtropical lowland humid
La Molina (LM), Lima	Spring-summer	Warm, semi-arid	12	133	Central coast	Yield, drought and heat tolerance, processing	Subtropical lowland arid
Tacna (Tac),	Summer	Hot, saline soils	18	500	Southern coast	Drought, soil salinity and adaptation to long days	Subtropical lowland arid
Huancayo, Junín	Winter–autumn	Cold	11	3200	Central highland	Yield potential, processing quality	Tropical highland cold
Oxapampa	Spring	Rainy, warm	10	1000	Central high jungle	Late blight	Tropical highland humid
Majes	Spring–summer	Warm, arid	16	1300	Southern coast	Drought tolerance, glycoalkaloid accumulation	Subtropical lowland arid

Trait Research in Support of the Agile Potato

Proximal sensing for efficient development of the agile potato: Remote sensing technologies are increasingly being used to assess crops by breeders and physiologists. These technologies have the potential to simplify evaluation of large breeding populations and improve selection efficiency over traditional field measurements. Different vegetative indices combined together may, for example, improve yield prediction (Rodriguez Junior et al. 2014). CIP's work in this area has aimed to explore genetic variation for the canopy temperature differential (CTD) and Normalized Difference Vegetation Index (NDVI) measurements and their relations with tuber yield components using 27 segregating families of the LTVR population under water stress, toward identifying families with tolerance to drought at early stages of selection. A high negative correlation (−0.66) of NDVI with Harvest index-fresh weight was observed, as were a high positive correlation (0.74) with plant wilting, a moderately high (0.50) correlation with plant vigor under drought conditions, and a significant negative correlation (−0.5) of CTD with tuber yield (tons/ha). Remote sensing techniques may provide a way to save resources and accelerate genetic gains through the early identification of families with tolerance to abiotic stress in a breeding program.

Trait research for understanding of complex, underground development: In breeding for a short vegetative cycle, CIP emphasizes bulking-based maturity, as opposed to foliar maturity or senescence. Tuber bulking results from two basic processes, tuber initiation and tuber growth (Radley et al. 1961). Like Andigenum Group potatoes, most varieties and clones from CIP's program yield significantly before physiological maturity, and development and harvest date cannot be assessed on foliar senescence alone. Indeed, if CIP-bred varieties are harvested when foliage has senesced, they will generally be well beyond the optimum time for harvest. Timing and duration of bulking vary with location, environmental factors, and genetics. Early tuber initiation and growth are necessary for acceptable production in areas or varieties in which potatoes are often harvested prior to physiological maturity.

To help provide early-maturing, resilient potatoes to farmers in medium and high latitudes of Asia, a panel of 162 CIP breeding lines was characterized for bulking behavior under four environmental conditions in field plots in La Molina, Peru. The environmental conditions for this study were: simulated long (16 h), and short (12 h) day lengths, in each of two seasons—cool winter and hot summer. The traits evaluated in the four field plots include yield (number and weight of tubers on two harvest dates) and dry matter content, in addition to which tuber induction was assessed by the cutting method of Ewing (1992). Tuber bulking ratio was calculated as the number of medium-sized and large tubers over the total number of tubers (tiny + small + medium + large) multiplied by the ratio of tuberized plants. Phenotypic assessment according to CIP's Protocol for tuber bulking maturity assessment of elite and advanced potato clones (https://research.cip.cgiar.org/potatoknowledge/bulking.php) enabled the classification of genotypes by maturity through cluster analysis according to Evanno and Regnaut (2005), and thereby, the identification of clones for which tuberization and bulking were more or less influenced by heat and photoperiod, including a small set of selections classified as day neutral.

Genotyping of the panel with version 1 of the SolCAP infinium array (8303 single nucleotide polymorphism (SNP) markers) enabled the identification of genomic regions and SNP markers influencing potato tuberization and bulking related traits affected by photoperiod and high temperatures. Prerequisites for this genome-wide association study (GWAS) included the calculation of the population membership matrix and SNP-based kinship relationship matrix and the estimation of linkage disequilibrium (LD decay to $r^2 = 0.2$ over 0.3 Mb) for the panel. GWAS-associated SNP jointly explained up to 60% of the phenotypic variation for the various traits, Therefore, MAS could be efficiently used for pre-screening of seedlings before further field evaluation, and contribute to shift the mean of the selection population for adaptation to warm, long day environments.

The same information was also used in a pilot study of genomic selection (GS). The underlying concept in GS is to model the entire complement of quantitative trait loci (QTL) effects across the genome to produce a genomic estimated breeding value (GEBV) from which progenitors or clones can be selected after genotyping. This pilot study tested several models proposed by Endelman (2011) to assess the feasibility of applying GS to improve selection of parental clones for early tuberization and bulking under long day-length. Prediction accuracies of ≥0.3 can be regarded as sufficiently high to merit implementation of GS (Heffner et al. 2010). Sufficiently high values were attained for tuberization induction, stolon length, bulking ratio at 75 and 90 days, and marketable tuber number at 75 days. As all markers are simultaneously fitted to capturing most of the trait heritability, GS can be expected to help reduce the length of selection cycles aimed at improving traits for which accurate phenotyping can be performed.

Prospects for True Hybrid Potato Breeding

Several countries and institutions have conducted research proposing the production of potato by sexual or botanical seed. Since the vegetative (tuber) planting material of clonally propagated potatoes is typically called seed potato, the term true potato seed (TPS) was introduced for tetraploid potato varieties produced from botanical seed. TPS aimed to overcome difficulties associated with production of vegetative seed including slow multiplication rate, accumulation of pathogens, and the bulky and perishable nature of tuber seed (Kidane-Mariam et al. 1985; Golmirzaie et al. 1994; Almekinders et al. 2009). As potato is a cross-fertilizing crop developed from heterozygous parents that do not "breed true," TPS is heterogeneous and thus variable for a large number of characteristics. While this variability is the source of variation for a clonal variety program, it is generally undesirable in varieties per se.

True hybrid potato breeding, on the other hand, aims at developing uniform true seed (TS) hybrid varieties from complementary superior inbred stocks developed at the diploid level. It would combine the advantages of TPS, (large numbers of disease-free propagules), diploid breeding (simplified genetics), and homogeneity (by way of combining inbred instead of heterozygous parents) in a new potato breeding system akin to that of maize.

Preliminary reports of productive diploid hybrids (de Vries et al. 2016), and the advantages of "re-inventing potato" as an inbred line-based crop (Jansky et al. 2016) make this effort one of the most intriguing developments in potato breeding over the past decade. Diploid inbreeding-tolerant potatoes would allow for considerably more systematic incorporation of new genes and traits, such as by backcrossing to fixed genotypes, as well as the possibility of substantial yield gains by crossing between well-defined heterotic groups. Working with diploid potatoes would also greatly facilitate the use of desirable recessive alleles. In principle, the low multiplication rate of clonally propagated potato can be circumvented by moving to diploid hybrids, where true seed from inbred × inbred crosses can produce vast numbers of genetically identical offspring in a short period of time.

Several steps need to be taken before diploid potato inbred line breeding might become routine. Most important is eliminating the self-incompatibility system, so that inbred lines can be readily developed. Recently, the dominant self-incompatibilty locus inhibitor allele (*Sli*) identified in certain self-compatible (SC) variants of the diploid wild potato *Solanum chacoense* (Hosaka and Hanneman 1998) has been used to overcome self-incompatibility (SI) and develop SC inbreds directly from the original source or by introgression to SI 2x cultivated germplasm (Phumichai and Hosaka 2006; Jansky et al. 2014). Additional research may be needed in order to manage SC as a trait in potato breeding, to improve the vigor of potato seedlings and to assure male fertility in diploid male parents.

Investment in inbreeding seems to be greater so far in the private than in the public sector. Nevertheless, some public breeding programs including CIP's have begun to seek some of the advantages that this novel strategy can offer to breeding and genetics for potato improvement. CIP has undertaken research to develop (1) a self-compatible) 2x breeding platform for broadening the genetic base of potatoes by hybridization across genepools, and (2) inbred genetic stocks for gene discovery and trait capture. Initial stocks and experience gained in the process of this research will contribute to longer term goals of developing complementary superior diploid inbred parental lines that will produce uniform, heterotic true seed 2x varieties when combined by crossing.

Variety Testing and Seed Links

How the Breeding Program Is Integrated with Seed Production in the State of New York in the United States

In addition to a public sector breeding program, the state of New York also maintains a public sector nuclear seed farm (the Uihlein Farm, in Lake Placid NY), which produces clean seed for commercial seed growers in the region. Each year the NY breeding program identifies up to five promising selections—typically clones that have been evaluated for at least 6 years in the field—and sends them to the Uihlein Farm. The Uihlein Farm first tests each clone for viruses, removing any that are present, and then, after a few years, depending on the multiplication scheme chosen, offers minitubers or limited-generation field grown tubers of these selections for sale.

Over time the breeding program has learned that it is important not to discuss promising clones with local industry until seed will soon be, or already is, available. If a clone elicits excitement, and growers ask for seed yet find none, they quickly lose interest, and it is very difficult to rekindle interest once seed does become available. When a clone is discussed before seed is available from the Uihlein Farm, the breeding program strives to provide up to 1000 kg of seed for commercial ware and chipstock growers to evaluate on their farms in the interim.

The NY seed farm bears some risk in this arrangement, as the breeding program typically discovers, over the course of additional years of evaluation, serious weaknesses in some of the clones being multiplied, rendering the clone unmarketable. On the other hand, the NY seed farm also benefits when it can be the first supplier of a new variety.

Regional Trials for Variety Assessment

Procedures, interfaces and responsibilities for trialing elite potato clones for variety release vary in degree of organization or formality across and sometimes within Countries.

As the aim of *plant breeding* is to produce new, improved *varieties, breeders should be aware of and anticipate meeting the several steps required for variety recognition, release and registration.*

Depending on program structure and target territory) and the complexity of the selection criteria, potato variety breeding typically involves a minimum of 6–7 years (and often more) of crossing and selection work, followed by a year or 2 of early testing to identify superior clones. The next step is the entry of candidate varieties into official variety registration recommendation trials. Programs in which breeding is done in the territory of release are more straight-forward in this respect than international programs in which additional time is required for quarantine exchange and bulking of seed in a new location. Depending on access to infrastructure and communication about objectives, this can take 2–5 years. The public and private sector differ in this respect since while public programs generally provide elite materials as true seed families or a small number of in vitro plants for research and breeding that may lead to local selection, breeding companies provide a reasonable quantity of seed of known varieties along with support and incentives for their testing and release.

Case Study 9: Regional Trials for Potato Variety Development in the United States

Northeast Regional Potato Trials—trialing system is a network for testing elite clones as candidate varieties. At present public-sector potato breeding in the US is divided into four regions—east, central, west and south. Each region has a regional trialing network, in which the agronomic performance of advanced clones is evaluated across a wide geographical range. The trial serves both to identify those relatively uncommon clones with broad adaptation, which are most likely to become commercially acceptable, and to identify varieties that may serve the needs of narrow geographic areas that have no breeding program of their own. Taking the eastern US as an example: the system consists of four public sector breeding programs—2 based in Maine, one in North Carolina, and 1 in New York—and 11 evaluation sites, 2 in NY and 1 each in Maine, North Carolina, Pennsylvania, Florida, Virginia, Ohio, Maryland and occasionally the Canadian provinces of New Brunswick and Prince Edward Island. Evaluators at each site rate clones using standardized scoring scales for yield, specific gravity, shape, internal and external defects, maturity, texture of tuber skin, and overall appearance, and upload the data to a common web portal (potatoes.ncsu.edu/NE.html). The originating breeding programs use the data to help make selection decisions, while regional evaluators use it to help inform local growers about which new varieties will perform best in their region. Breeders and evaluators also meet once a year to discuss, in person, the performance of clones in the trial the preceding year, and to make incremental adjustments in how the trial data is collected, analyzed and reported. Over the past decade separate national programs in the US, focused on chip and french fry variety development, have built upon the standardized framework of the eastern region trial network, with that data housed at potatoesusa.mediusag.com.

How Does the New York Program Decide When to Release a Variety?

In many countries a new potato variety cannot be legally grown until it has been formally evaluated in registration trials, found to be distinct, uniform and stable (DUS), and shown to have "value for cultivation and use." Decisions on all of these characters are made by an independent regulatory body, not the originating breeder.

In contrast, in the United States there is no legal requirement for comparable tests before a potato variety can be grown. (Although there is a requirement for DUS data, if a US breeder wants to procure plant breeders rights.) While some US universities have panels of breeders that evaluate candidate varieties, and can deny release of a variety by a colleague, others, like Cornell University in New York, have no formal potato variety release process at all.

Is either of these approaches preferable? Coauthor Walter de Jong is comfortable to operate in an unregulated environment. In his view the uniformity and stability tests are essentially meaningless for a clonally propagated crop. In addition, while he recognizes that evaluating distinctness is important to prevent anyone from re-releasing varieties developed by others, as long as two highly heterozygous autotetraploids have been crossed, all offspring will be distinct, as so many loci segregate. In other words, it is trivial to create a distinct potato. Variety release decisions in New York are made, instead, by gauging stakeholder interest. Prior to formal release, seed growers are permitted to grow candidate quantities under an evaluation license. When a candidate variety reaches a threshold, typically 10 ha or more of seed across all seed growers, and the seed growers ask for the variety to receive a formal (nonexperimental) name, it is released. One specific situation where the New York program will not release a variety, even if there is industry interest, is if tuber glycoalkaloids are too high (>20 mg/100 g tuber fresh weight).

Case Study 10: All India Coordinated Crop Improvement Project and Release of Kufri-Lima from LTVR Population

In India, state and national systems for verity testing and release coexist, and to some degree depend on each other, and these policies serve as a model for other countries in south Asia. Formal testing of a candidate variety by the All India Coordinated Crop Improvement Projects (AICCIP), and the subsequent recommendation of the variety for release by the AICCIP are required before variety can be released by the Central Sub-Committee that permits notification and thereby enables seed production and trade. In the case of potato, a minimum of 3 years of multilocational trials and assessment for Value for Cultivation and Use (VCU) take advantage of 16 locations across the country. These locations use standardized design and data templates and report on performance as well as meteorological conditions. Experiments are designed for Rabi (winter) season for irrigated areas such as Central, Northern, and Eastern regions (Indo-Gangetic Plains). Plateau regions (700–1200 masl) and hilly regions (1500–3000 masl) are represented in the multilocational trial series.

A new variety should be suitable for specified agro-climatic and soil conditions, have an ability to withstand typical stress conditions, and have tolerance/resistance to pests and diseases. It should also show distinct advantages over the existing equivalent released varieties, a process facilitated by standard formats for data collection. The breeders or Principal Investigators and Zonal Co-ordinators attend the meeting to provide wider information on the variety. The Director of Crop Development Programme is invited to provide information on the response of farmers to promotion and demonstration trials if they were conducted. The release proposal proforma requires the breeder to ensure availability of enough seed stock for seed multiplication on at least 10 ha.

An interesting feature of the All India trials project is that it reports on potato varieties as components of cropping systems, and not only on potato as a commodity. The pertinent proforma includes information on the productivity and profitability of potato and each of different crops in intercropping systems. Long-term experiments address issues of nutrient cycling and crop management that offer valuable methodology and possibly valuable data sets and recommendations for CIP's concern for the resilience of its Agile Potato proposition, proposing intensification of cereal-based systems with attention to sustainability.

Recent emphasis of collaboration between CIP and the Central Potato Research Institute (CPRI) of India have been on identifying early-maturing, heat tolerant, virus-resistant varieties suitable for intensifying and diversifying rice-based production systems. CIP introduces elite clones as in vitro plants as well as true seed from crosses between parents with good GCA for features of performance in subtropical lowland environments for used in breeding or direct identification of varieties for the vast Indo-Gangetic Plains and hilly regions of southwest Asia. The release of Kufri-Lima in October 2017, bred as CIP397065.28, is an exciting result of this collaboration. Potato virus are a serious problem for farmers in India, particularly in Gujarat, Rajasthan, Madhya Pradesh, Chhattisgarh, Odisha, West Bengal, and East Uttar Pradesh, where temperatures are higher in the winter season. Crop losses due to viruses range between 20 and 50%, and farmers from these areas traditionally need to import seed from Northern India (Punjab). "Farmers usually have to wait until temperatures drop to plant potatoes, but because of its tolerance to heat, Kufri Lima can be planted a full 20–30 days earlier than other local varieties." Earlier planting means earlier harvests, giving Kufri Lima farmers the ability to sell their potatoes at a premium price before other varieties hit the market. Farmers can expect to be paid 40–50% higher prices than those who harvest potatoes during the normal season. Heat tolerance and early maturity mean that Kufri Lima farmers can invest their earnings into a second winter harvest helping them to improve the overall productivity of the cropping system, while virus resistance may enable seed production closer to home. https://www.potatopro.com/companies/central-potato-research-institute-cpri.

Kufri-Lima was introduced to CPRI as CIP397065.28 in 2006 after initial selection in Peru. It was evaluated for adaptation, quality, and resistance to degeneration along with several other selections from the LTVR population with the support of ICAR (International Council of Agricultural Research), Central Potato Research Institute, GIZ/BMZ and USAID.

Variety Release and Registration

Keen awareness of timelines, procedures and actors involved in decisions on the release of varieties from elite clones and of policies and practices that permit or enable seed production is essential for the uptake of new varieties and subsequent benefit of their component and composite traits. Country specific systems for variety testing, release and dissemination of seed to farmers are governed by sets of national laws, scientific guidelines, norms, and standard practices which together can be termed "Regulatory Frameworks."

Registration is often required for seed to be multiplied and marketed in a country or territory or for exportation outside the territory. Release is intended to make a newly developed variety available to the public for general cultivation in the regions for which it is adapted. Even though they are not directly responsible, breeders should be aware of the designated service responsible for recommending the registration of new varieties in their target locations, as well as the Basic information required for application. Additional efficiencies can be realized when breeding programs are aware of and contribute to the availability of Basic material required for application and variety release. Knowledge of the agency with which a reference sample has to be deposited, up to date reference to the format for information required and the quantity of seed to be available at time of release vary be country or state. Depending on the territory in question, the designated service responsible for approving the registration of new varieties is usually a ministry or department of agriculture, which may have crop specific rules and regulations for variety development. In India and Bangladesh, for example, potato is among the crops for which a special category called "notified" is required for varieties to enter seed production schemes.

Additional policies, processes and agencies are those responsible for demonstration or promotion trials for pre-released and released varieties where required; maintenance of certification standards; and the production and storage of foundation seed and provision of seed of new varieties. Policy regulating seed trade is of less direct concern, though also important to breeding programs. An excellent reference for procedures and policy for variety testing and release can be found at http://www.coraf.org/wasp2013/wp-content/uploads/2013/07/ECOWAS_VAR_REGIST_MANUAL_SEP_081.pdf.

Regional Frameworks for Variety Release

Beyond national seed regulatory systems, several regional frameworks are being developed or implemented across countries. Advantages of regional frameworks include improved access by farmers to seed of improved varieties at affordable prices, avoidance of repetitive national testing, making seed trade easier, faster, and less expensive for introduction of new varieties, timely availability of quantity and high quality of seed and choice of variety needed. Regional frameworks encourage individual countries to adopt national seed legislation and establish national variety catalogues that contribute to growers' awareness of new varieties. Regional registration can result in a larger market share for varieties and reduced time and cost for national variety release when second and third country releases are simplified by admissibility of available performance trial data.

Fast track options for variety registration at regional level rely on policies and communication practices that recognize prior registration in one member country and one season testing as sufficient for variety release in a second one. Similarly relevant are federal and state or provincial procedures that complement each other, enabling the benefits of new varieties to be realized across states or countries. Mechanisms are also needed whereby release proposal on the basis of data from state and farmers' field trials could be used for the zonal release of a variety, across an agro-ecological zone that covers more than one state or one country (Tripp et al. 1997). These same authors conclude that adjustment to seed regulatory frameworks is necessary because of significant changes in national seed systems, including: reductions in budget for public agricultural research; the failure of many seed parastatals; increasing concern about plant genetic diversity; pressure for the establishment of plant variety protection; the increasing contributions of commercial seed enterprises; and the emergence of innovative local level variety development and seed production initiatives.

A report of test cases in Africa supported by Syngenta documented the inclusion of three new potato varieties in the COMESA (Common Market for Eastern & Southern Africa) regional catalogue in 2016 (https://www.syngentafoundation.org/

sites/g/files/zhg576/f/seeds_policy_regional_variety_release_test_cases_317.pdf). Notably, these are the first non-maize varieties to be included in the region's list.

Concluding Remarks

Potato breeding is at a crossroad. Gene editing and large scale genotyping are at hand, and true hybrid breeding appears feasible, even for this complex polyploid and traditionally clonally propagated crop. Together with pre- and post-Mendelian methods, these advances promise to help accelerate genetic gain to nourish a growing population, help preserve the environment, and confront economic and climate constraints. Sustained breeding and supporting trait research will assure potato's role in a more comprehensive and integrated agriculture in which quality and resilience, in addition to the historical emphasis on yield, are major goals. Community efforts including the development of cross-species tools and research aimed at understanding the principles underlying crop performance will be needed to meet growing demands on agriculture. The programs illustrated here use crop ontologies to support understanding of traits and attempt to exploit heterosis, inter-ploidy breeding, and accelerated selection schemes to reach their goals. Beyond this, research on root systems, G × E interaction and photosynthetic efficiency merit attention. The lengthy 12–20 years required to develop and release a new potato variety, coupled with climate variability, call for intensified research on selection strategies and choice of breeding method. The use of marker-assisted and genomic selection, the possibilities offered by inbreeding coupled with gametophyte selection and hybrid development, "big data" and efficient real-time phenotypic data collection and analysis are most compelling at present. Finally, collaboration across sites and disciplines and the utilization of statistical models for assessing genotypic adaptation and breeding value are critical to the efficient development and deployment of better potato varieties.

Acknowledgements We gratefully acknowledge Elisa Mihovilovich and Thiago Mendes for valuable discussions toward this chapter.

References

Almekinders CJM, Chujoy E, Thiele G (2009) The use of true potato seed as pro-poor technology: the efforts of an international agricultural research institute to innovating potato production. Potato Res 52:275–293

Bae J, Halterman D, Jansky S (2008) Development of a molecular marker associated with Verticillium wilt resistance in diploid interspecific potato hybrids. Mol Breed 22:61–69

Bernardo R (2002) Breeding for quantitative traits in plants. Stemma Press, Woodbury

Bonierbale M, Plaisted RL, Tanksley S (1993) A test of the maximum heterozygosity hypothesis using molecular markers in tetraploid potatoes. Theor Appl Genet 86:481–491

Bonierbale M, Amoros W, Burgos G, Salas E, Suarez H (2007) Prospects for enhancing the nutritional value of potato by plant breeding. African Potato Association conference proceedings, Alexandria, Egypt, vol 7, pp 26–46

Butler DG, Cullis BR, Gilmour AR, Gogel BJ (2009) ASReml-R reference manual. v. 3. Queensland Department of Primary Industries & Fisheries, Brisbane

Chase SS (1963) Analytic breeding in *Solanum tuberosum* L.—a scheme utilizing parthenotes and other diploid stocks. Can J Genet Cytol 5:359–363

Colton L, Groza H, Wielgus S, Jiang J (2006) Marker-assisted selection for the broad-spectrum potato late blight resistance conferred by gene RB derived from a wild potato species. Crop Sci 46:589–594

Comai L (2005) The advantages and disadvantages of being polyploid. Nat Rev Genet 6:836–846

Comstock RE, Robinson HF (1948) The components of genetic variance in populations of biparental progenies and their use in estimating the average degree of dominance. Biometrics 4:254–266

Cubillos A, Plaisted R (1976) Heterosis for yield in hibrids between *S. tuberosum* spp. *Tuberosum* and *S. tuberosum* spp. *Andigena*. Am Potato J 53:143–150

De Jong H, Rowe PR (1971) Inbreeding in cultivated diploid potatoes. Potato Res 14:74–83

De Nettancourt D (1977) Incompatibility in angiosperms. Springer, Berlin

Endelman JB (2011) Ridge regression and other kernels for genomic selection with R package rrBLUP. Plant Genome 4:250–255

Evanno G, Regnaut S (2005) Detecting the number of clusters of individuals using the software STRUCTURE: a simulation study. Mol Ecol 14:2611–2620. https://doi.org/10.1111/j.1365-294X.2005.02553.x

Ewing EE (1992) Critical photoperiods for tuberization: a screening technique with potato cuttings. Am Potato J 55:43–53

Finkers-Tomczak A, Bakker E, de Boer J, van der Vossen E, Achenbach U, Golas T, Suryaningrat S, Smant G, Bakker J, Goverse A (2011) Comparative sequence analysis of the potato cyst nematode resistance to locus H1 reveals a major lack of co-linearity between three haplotypes in potato (*Solanum tuberosum* ssp.). Theor Appl Genet 122:595–608

Frankel R, Galun E (1977) Pollination mechanisms, reproduction, and plant breeding. Springer, New York

Fuglie K (2007) Priorities for potato research in developing countries: results of a survey. Am J Potato Res 84:353–365

Gaiero P, Mazzella C, Vilaró F, Speranza P, de Jong H (2017) Pairing analysis and in situ hibridisation reveal autopolyploid-like behaviour in *Solanum commersonii* × *S. tuberosum* (potato) interspecific hybrids. Euphytica 213:137. https://doi.org/10.1007/s10681-017-1922-4

Gallais A (2004) Quantitative genetics and selection theory in autopolyploid plants. INRA, Paris

Gálvez R, Mendoza HA, Fernández-Northcote EN (1992) Herencia de la inmunidad al virus Y de la papa (PVY) en clones derivados de *Solanum tuberosum* spp. Andigena. Fitopatología 27:8–15

Gardner CO, Heberhart SA (1966) Analysis and interpretation of the variety cross diallel and related populations. Biometrics 22:439–452

Gatto M, Hareau G, Pradel W, Suarez V, Qin J (2018) Release and adoption of improved potato varieties in Southeast and South Asia. International Potato Center (CIP) Lima, Peru, 42p. Social Sciences Working Paper No. 2018-2. ISBN 978-92-9060-501-0

Gebhardt C, Bellin D, Henselewski H, Lehmann W, Schwarzfischer J, Valkonen J (2006) Marker-assisted combination of major genes for pathogen resistance in potato. Theor Appl Genet 112:1458–1464

Golmirzaie A, Malagamba P, Pallais N (1994) Breeding potatoes based on true seed propagation. In: Bradshaw J, MacKay G (eds) Potato genetics. CAB International, Wallingford, pp 499–512

Griffing B (1956) Concepts of general and specific combining ability to relation to diallel crossing system. Anst J Biol Sci 9:463–493

Grüneberg WJ, Mwanga R, Andrade M, Espinoza J (2009) Chapter 5: Selection methods Part 5: Breeding clonally propagated crops. In: Ceccarelli S, Guimarães EP, Weltzien E (eds) Plant breeding and farmer participation. FAO, Rome, pp 275–322

Hareau G, Kleinwechter U, Pradel W, Suarez V, Okello J, Vikraman S (2014) Strategic assessment of research priorities for potato. CGIAR Research Program on Roots, Tubers and Bananas (RTB), Lima (Peru). RTB Working Paper 2014-8. Available online at: www.rtb.cgiar.org

Harlan IR, de Wet IMI (1971) A rational classification of cultivated plants. Taxon 20(4):509–517

Hayman BI (1954) The theory and analysis of diallel crosses. Genetics 39:789–809

Heffner EL, Lorenz AJ, Jannink J l, Sorrells ME (2010) Plant breeding with genomic selection: Gain per unit time and cost. Crop Sci 50(5):1681–1690

Hirsch C, Hisrch CD, Felcher K, Coombs J, Zarka D, VanDeynze A, DeJong W, Veilleux RE, Jansky S, Bethke P, Douches DS, Buell CR (2013) Retrospective view of North American potato (*Solanum tuberosum* L.) breeding in the 20th and 21st centuries. G3 Genes Genomes. Genetics 3:1003–1013

Hirsch CD, Hamilton J, Childs K, Cepela J, Crisovan E, Vaillancourt B, Hirsch CN, Habermann M, Neal B, Buell CR (2014) Spud DB: a resource for mining sequences, genotypes, and phenotypes to accelerate potato breeding. Plant Genome 7(1):1–12

Hosaka K, Hanneman RE (1998) Genetics of self-compatibility in a self-incompatible wild diploid potato species *Solanum chacoense*. 1. Detection of an S locus inhibitor (Sli) gene. Euphytica 99:191–197. https://doi.org/10.1023/A:1018353613431

Jansky SH, Chung YS, Kittipadukal P (2014) M6: a diploid potato inbred line for use in breeding and genetics research. J Plant Regist 8:195–199. https://doi.org/10.3198/jpr2013.05.0024crg

Jansky SH, Charkowski AO, Douches DS, Gusmini G, Richael C, Bethke PC, Spooner DM, Novy RG, De Jong H, De Jong WS, Bamberg JB, Thompson AL, Bizimungu B, Holm DG, Brown CR, Haynes KG, Sathuvalli VR, Veilleux RE, Miller JC, Bradeen JM, Jiang JM (2016) Reinventing potato as a diploid inbred line–based crop. Crop Sci 56:1–11

Kasai K, Morikawa Y, Sorri VA, Valkonen JP, Gebhardt C, Watanabe KN (2000) Development of SCAR markers to the PVY resistance gene Ryadg based on a common feature of plant disease resistance genes. Genome 43:1–8

Kempthorne O, Curnow RN (1961) The partial diallel cross. Biometrics 17:229–250

Kidane-Mariam HM, Mendoza H, Wissar R (1985) Performances of true potato seed families derived from intermating tetraploid parental lines. Am Potato J 62:643–652. https://doi.org/10.1007/BF02853473

Kleinwechter U, Hareau G, Suarez V (2014) Prioritization of options for potato research for development—results from a global expert survey. CGIAR Research Program on Roots, Tubers and Bananas (RTB), Lima (Peru). RTB Working Paper 2014-7. Available online at: www.rtb.cgiar.org\

Levin DA (1983) Polyploidy and novelty in flowering plants. Am Nat 122:1–25

Lynch M, Walsh B (1998) Genetics and analysis of quantitative traits. Sinauer Associates, Sunderland

Meenakshi JV, Johnson N, Manyong VM, De Groote H, Javelosa J, Yanggen D, Naher F, Gonzalez C, Garcia J, Meng E (2007). How cost-effective is biofortification in combating micronutrient malnutrition? An ex-ante assessment HarvestPlus Working Paper No.2/. HarvestPlus, Washington, DC. www.harvestplus.org

Mendiburu AO, Peloquin SJ (1977) Bilateral sexual polyploidization in potatoes. Euphytica 26:573–583

Mendoza HA (1990) Mejoramiento poblacional. Una estrategia para la utilización del germoplasma de papa. En: Mejoramiento de papa en el Cono Sur, pp 47–62

Mendoza HA, Estrada NR (1979) Breeding potatoes for tolerance to stress: heat and frost. In: Mussel H, Staples RR (eds) Stress physiology in crop plants. Wiley, New York, pp 227–262

Mendoza HA, Sawyer RL (1985) The breeding program at the International Potato Center (CIP). In: Russell GE (ed). Progress in plant breeding, 1. London, Butterworths, 325 p

Mendoza HA, Mihovilovich EJ, Saguma F (1996) Identification of triplex (YYY) potato virus Y (PVY) immune progenitor derived from *Solanum tuberosum* spp. Andigena. Am Potato J 73(1):13–19. ISSN 0003-0589

Midmore DJ, Rhoades RE (1988) Applications of agrometeorology to the production of potato in the warm tropics. In: Rijks D, Stigter CJ (eds) Agrometeorology of the potato crop. International Society for Horticultural Science (ISHS); International Potato Center (CIP); World Meteorological Organization (WMO); European Association for Potato Research; Royal Netherlands Meteorological Institute, Wageningen, pp 103–136. Acta Horticulturae. no. 214

Mihovilovich E, Alarcón L, Pérez AL, Alvarado J, Arellano C, Bonierbale M (2007) High levels of heritable resistance to potato leafroll virus (PLRV) in *Solanum tuberosum* subsp. Andigena. Crop Sci 47(3):1091–1103

Mihovilovich E, Aponte M, Linqvist-Kreuze H, Bonierbale M (2014) An RGA-derived SCAR marker linked to PLRV resistance from *Solanum tuberosum* ssp. Andigena. Plant Mol Biol Report 32:117–128. https://doi.org/10.1007/s11105-013-0629-5

Muñoz FJ, Plaisted RL, Thurston HD (1975) Resistance to potato virus Y in *S. tuberosum* spp. Andigena. Am Portao J 52:107–115

Oakey H, Verbyla AP, Cullis BR, Wei XM, Pitchford WS (2007) Joint modeling of additive and non-additive (genetic line) effects in multi-environment trials. Theor Appl Genet 114:1319–1332. https://doi.org/10.1007/s00122-007-0515-3

Ortiz R (1998) Potato breeding via ploidy manipulations. Plant Breed Rev 16:15–85

Paget M, Amoros W, Salas E, Eyzaguirre R, Alspach P, Apiolaza L, Noble A, Bonierbale M (2014) Genetic evaluation of micronutrient traits in diploid potato from a base population of Andean Landrace Cultivars. Crop Sci 54:1949–1959

Phumichai C, Hosaka K (2006) Cryptic improvement for fertility by continuous selfing of diploid potatoes using Sli gene. Euphytica 149:251–258. https://doi.org/10.1007/s10681-005-9072-5

Piepho H-P, Möhring J, Melchinger AE, Büchse A (2008) Blup for phenotypic selection in plant breeding. Euphytica 161:209–222

Plaisted RL (1987) Advances and limitations in the utilization of Neotuberosum in potato breeding. In: Jellis GJ, Richardson DE (eds) The production of new potato varieties. Cambridge University Press, Cambridge, pp 186–196

Plaisted RL, Sanford L, Federer WT, Kehr AE, Peterson LC (1962) Specific and general combining ability for yield in potatoes. Am Potato J 39:185–197

Pushkarnath P (1942) Studies on sterility in potatoes. 1. The genetics of self- and cross-incompatibilities. Indian J Genet and Pl Br 2:11–36

Radley RW, Taha MA, Bremner PM (1961) Tuber bulking in the potato crop. Nature 191:782–783

Ragot M, Bonierbale M, Weltzien E (2018) From market demand to breeding decisions—a framework. RTB Working Paper.

Rodriguez Junior FA, Ortiz-Monasterio I, Zarco-Tejada PJ, Ammar K, Gérard B (2014) Using precision agriculture and remote sensing techniques to improve genotype selection in a breeding program. Conference Paper.International Conference on Precision Agriculture. http://quantalab.ias.csic.es/pdf/Rodrigues%20Jr.%20et%20al.%202014.pdf

Salas E (2002) Estudio de la variabilidad genética de la resistencia al Virus del Enrollamiento de la papa (PLRV) en una población tetraploide. Tesis para optar el Grado de Magister Scientiae. Universidad Nacional Agraria—La Molina. Lima, Peru

Sanford JC, Hanneman RE (1982) Intermating of potato haploids and spontaneous sexual polyploidization—effects on heterozygosity. Am J Potato Res 59:407–414

Sattarzadeh A, Achenback U, Lubeck J, Strahwald J, Tacke E, Hofferbvert H, Rothsteyn T, Gebhardt C (2006) Single nucleotide polymorphism (SNP) genotyping as basis for developing a PCR-based marker highly diagnostic for potato varieties with high resistance to Globodera pallida pathotype Pa2/3. Mol Breed 18:301–312

Slater AT, Wilson GM, Cogan NO, Forster JW, Hayes BJ (2014) Improving the analysis of low heritability complex traits for enhanced genetic gain in potato. Theor Appl Genet 127(4):809–820

Spooner DM, Ghislain M, Simon R, Jansky SH, Gavrilenko T (2014) Systematics, diversity, genetics, and evolution of wild and cultivated potatoes. Bot Rev 80:283–383. https://doi.org/10.1007/s12229-014-9146. —Toward a Rational Classification of Cultivated Plants—Jstor https://www.jstor.org/stable/1218252

Sprague GF, Tatum A (1941) General vs. specific combining ability in single crosses of corn. J Am Soc Agron 34:923–932

Tripp R, Louwaars NP, Van Der Burg WJ, Virk DS, Witcombe JR (1997) Alternatives for Seed regulatory reform: an analysis of variety testing, variety regulation and seed quality control, Agricultural Research and Extension Network Paper, no. 69. http://hdl.handle.net/10535/4581

Ugent DL, Pozorski S, Pozorski T (1982) Archaeological potato tuber remains from the Casma Valley of Peru. Econ Bot 36:182–192

Ugent DL, Dillehay T, Ramirez C (1987) Potato remains from a Late Pleistocene settlement in southcentral Chile. Econ Bot 41:17–27

Uitdewilligen JGAML, Wolters A-MA, D'hoop BB, Borm TJA, Visser RGF, van Eck HJ (2013) A next-generation sequencing method for genotyping-by-sequencing of highly heterozygous autotetraploid potato. PLoS One 8(5):e62355. https://doi.org/10.1371/journal.pone.0062355

Velásquez AC, Mihovilovich E, Bonierbale M (2007) Genetic characterization and mapping of major gene resistance to potato leafroll virus in *Solanum tuberosum* ssp. *Andigena*. Theor Appl Genet 114(6):1051–1058

De Vries MI, ter Maat M, Pim Lindhout P (2016) The potential of hybrid potato for East-Africa. Open Agric. 1: 151–156

Wricke G, Weber WE (1986) Quantitative genetics and selection in plant breeding. de Gruyter, Berlin

Xu X et al (2011) Genome sequence and analysis of the tuber crop potato. Nature 475(7355):189–195

Yuen J, Forbes G (2009) Estimating the level of susceptibility to *Phytophthora infestans* in potato genotypes. The American Phytopathological Society. Phytopathology 99(6):2

Zhang LH, Mojtahedi H, Kuang H, Baker B, Brown CR (2007) Marker-assisted selection of Columbia root-knot nematode resistance introgressed from *Solanum bulbocastanum*. Crop Sci 47:2021–2026

2

Ex Situ Conservation of Potato [*Solanum* Section *Petota* (Solanaceae)] Genetic Resources in Genebanks

David Ellis, Alberto Salas, Oswaldo Chavez, Rene Gomez, and Noelle Anglin

Abstract Conserving the genetic diversity of potato is critical for the long-term future of potato improvement programs. Further, it is the social and ethical responsibility of the present generation to ensure future generations have the same opportunities to use, exploit, and benefit from the genetic diversity that exists today. Genebanks and the ex situ conservation of potato genetic resources are the only way to ensure this happens; in situ conservation plays a complementary role, but it can never ensure that the vast diversity that exists on earth today is still there for use in the future. Material in ex situ genebanks not only serve as a reservoir of ready-to-use genetic material when needed but also provide invaluable tools for research now and in the future of cultivated potato and its wild relatives.

Ex Situ Conservation of Potato

The conservation of crop diversity, such as potato diversity, outside of its natural habitat, or ex situ, is held in botanical gardens and genebanks throughout the world. While botanical gardens conserve diversity and display this diversity for the public to enjoy and learn from, as well for scientific research, the overall mandate of botanical gardens is not generally to promote the use of and share the plant diversity they hold. In contrast, the mandate of genebanks *is* to provide access to the genetic resources they hold and to promote the use of these genetic resources for training, breeding, and research. Unfortunately, many national genebanks holding potato germplasm do not have the capacity, resources, nor a national mandate to widely distribute their holdings, and thus, access to potato germplasm is mostly left to the genebanks in developing countries, United States Department of Agriculture (USDA), and the potato genebank of the CGIAR, which holds potato germplasm at the International Potato Center, or CIP after its Spanish acronym located in Lima, Peru.

D. Ellis (✉) · A. Salas · O. Chavez · R. Gomez · N. Anglin
International Potato Center, Lima, Peru
e-mail: d.ellis@cgiar.org

Potato genetic resources held ex situ are divided into potato wild relatives (maintained as seed populations) and cultivated varieties which are maintained either in field plantings (where tubers are planted and harvested annually), seed, in vitro/clonal material, or cryopreserved material. The same genetic material could be held in two or more different forms in a single genebank (in vitro and cryopreserved) or between two genebanks (seed and in vitro). What is critical for a genebank, is that the material it holds is available in a form that can be readily used by all who need it, including breeders, large and small-holder farmers, industry, researchers from developing or developed countries, and for teaching and educational purposes. Thus, the wealth of potato diversity held and made available by genebanks is an asset for humanity, as it safeguards known genetic traits which are useful today, along with a myriad of known and unknown traits ready to be deployed to meet the needs for the threats and challenges potato farmers face in the future.

The more uniform the crop in general, the easier it is to manage as an ex situ collection. Unfortunately, cultivated potato is an incredibly diverse crop, and while virtually all the potato planted and used commercially worldwide are tetraploid potato varieties, diploid, triploid, tetraploid, and even pentaploid landrace potatoes are still grown throughout the South American Andes. This diversity has been collected, maintained and is available for use in breeding and research programs worldwide from numerous genebanks. The diversity of cultivated potato is reflected not only in its vast ploidy range but also at the species level where over 30 years ago cultivated

Fig. Diversity of flower color and shape in cultivated potato from CIP germplasm collection

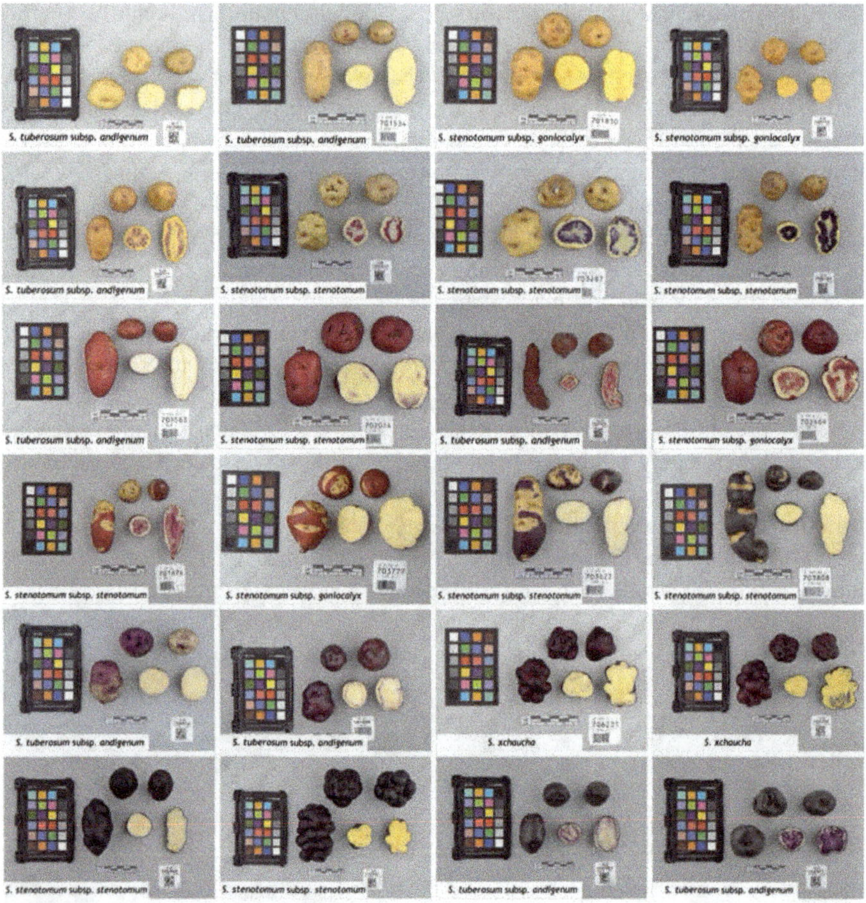

Fig. Diversity of tuber shape and morphology and skin and flesh color in cultivated potato from the CIP germplasm collection. The color card in each photograph is used for standardizing measurement of colors of the tubers

potato was divided into seven species and nine taxa [*Solanum tuberosum* (C. Linneo) *subsp. andigenum* (Juz. & Bukasov), *S. tuberosum* (C. Linneo) subsp. *tuberosum, S. xchaucha* (Juz & Bukasov), *S. stenotomum* subsp. *goniocalyx* (Juz. & Bukasov), *S. stenotomum* subsp. *stenotomum* (Juz. & Bukasov), *S. xjuzepczukii* (Bukasov), *S. phureja* (Juz. & Bukasov), *S. xajanhuiri* (Juz. & Bukasov), and *S. xcurtilobum* (Juz. & Bukasov); (Hawkes 1990)]. More recently the taxonomy of potato has been revised and cultivated potato has been regrouped into four species (*Solanum tuberosum, S. xjuzepczukii*, *S. xajanhuiri*, and *S. xcurtilobum*; Spooner et al. 2014). A similar taxonomic revision has occurred in the taxa of potato wild relatives with the previous taxonomic classification of Hawkes (1990) recognizing 228 wild species in 21 series. However, the recent taxonomic classification by Spooner et al. (2014) reduced the wild relatives of potato (*Solanum* section *Petota*) to 107 species.

The change in taxonomy has created a split in the ex situ potato world with some genebanks (such as the International Potato Center (CIP)] still organizing their collection based on the taxonomy of Hawkes (1990) while other major potato genebanks [such as the United States Department of Agriculture-Agricultural Research Service (USDA-ARS)] adopting the taxonomy of Spooner et al. (2014). Potato curators in all genebanks acknowledge the revision by Spooner et al. (2014) as needed and an advance in the field, yet the groupings by Hawkes (1990) make sense for managing an ex situ collection. Spooner et al. (2014) foresaw such challenges in their revision of the potato taxonomy and therefore stated that due to the "*complicating biological factors in section Petota*", they consider their "*taxonomy to be subject to critique and modification*" (Spooner et al. 2014, p 325). Two final comments are that there is consensus that ploidy in potato is not a good factor in delimitating species boundaries and that the use of different taxonomic names for the same material in different major potato genebanks is a detriment to use of the materials for the potato research community. That said, harmonizing the taxonomy used in global potato genebanks will need to be a focus in the next decade along with an extensive genetic comparison of potato collections to identify unique and redundant material between genebanks.

Taxonomic differences aside, ex situ collections and genebanks continue to play an ever increasingly dynamic role in providing the tools needed for food security in the future. The Second Report on The State of the World's Plant Genetic Resources for Food and Agriculture (FAO 2010) estimates that there are over 1750 genebanks worldwide holding approximately 7.4 million accessions; however, the report estimates that only 25–30% of these accessions are genetically unique. The report lists 98,285 potato accessions conserved in 174 genebanks around the world, and using the estimate above of the percentage of unique accessions, we can assume that there are at most an estimated 24,500–29,500 *unique* potato accessions are conserved worldwide in genebanks. In the case of potato, the accuracy of such an estimate of unique accessions is difficult to access until a thorough rationalization-comparison amongst the potato ex situ collections is carried out. According to this same report, 6 genebanks hold 41% of the global potato accessions: The French National Institute for Agricultural Research (INRA) in France (11%), Vavilov Institute in Russia (9%), The International Potato Center (CIP) in Peru (8%), The Leibniz Institute of Plant Genetics and Crop Plant Research (IPK) in Germany (5%), USDA-ARS in USA (5%), and The National Institute of Agrobiological Sciences (NIAS) in Japan (3%) and 20 global genebanks hold over 1000 potato accessions each. The listed global potato holdings collectively consist of 15% wild relatives of potato, 20% cultivated potato, 16% research and breeding materials, 14% advanced breeding lines, and 35% uncategorized accessions in these genebanks.

The last Global Strategy for Ex Situ Conservation of Potato (2006) analyzed 23 global potato collections which collectively maintain "*nearly 59,000*" potato accessions. This summary states that the genebanks in Latin America contain principally native cultivars while those in Europe and North America contain modern cultivars, breeding materials, and wild relatives. Such generalizations oversimplify the collections and indeed, although there is specialization among the potato genebanks, most major collections have a good representation of both wild and cultivated accessions.

Information about the particular attributes for each individual accession is critical for users to select which one or two of the thousands of accessions contain the trait(s) the user needs. Unfortunately, the information on accessions in global potato genebanks is generally incomplete and lacking the trait information users want and need most. Further, information on qualitative, complex traits (i.e., drought, frost tolerance, or yield) can be very specific to the physiological age at which the plant is subjected to the stress, the location, soil type, and degree of stress and, thus, is not always easy to list in a genebank database. Further, these complex traits require multiyear, multilocation evaluations in order to understand the complexities of the traits and genes associated with the trait. The USDA-ARS collection contains a large list of descriptors and traits recorded with 137 morphological, biochemical, nutritional, or physiological traits listed for potato (https://npgsweb.ars-grin.gov/gringlobal/cropdetail.aspx?type=descriptor&id=73), while CIP lists 72 descriptors or traits for their potato holdings (http://genebank.cipotato.org/gringlobal/search.aspx) and the European Union Genetic Resources of Potato lists 58 descriptors or traits (http://ecpgr.cgn.wur.nl/eupotato/). All these data are accessible in English on public internet-accessible databases, as is information from other global potato genebanks; although far too much of the information of the potato ex situ collections is not available publicly and outside of a few genebanks, much of the available listings from genebanks contain little information about the accessions. Fortunately, efforts such as Genesys, the global Gateway to Genetic Resources (https://www.genesys-pgr.org/welcome), provides descriptor and trait information from multiple genebanks in a single database, which facilitates use by facilitating the location and finding information on potato accessions worldwide.

Collection of New Potato Germplasm

There are estimations that 20% of plant species are in danger of extinction (Jansky et al. 2013) and such losses will be accelerated by climate change and habitat destruction. Further, for some plants, a large amount of their diversity may already only be able available from genebanks (Jarvis et al. 2011), which might be the future for potato where the wild relative species exist only in the fragile ecosystems of the Andes (Hijmans and Spooner 2001), where a changing climate is already having a substantial impact. Lingering questions for ex situ and in situ management of potato revolve around how to define diversity there is/was in cultivated and the large secondary genepools (wild potato relatives) of potato. For conservationists, the big question is how much of the diversity in potato is securely conserved, what diversity has been lost or is in imminent danger of being lost, and what is the economic value, as well as potential future value, of the potato diversity that is not securely conserved.

A benchmark study (Castañeda-Álvarez et al. 2015) evaluated the status of ex situ collections for 73 species (according to Spooner et al. 2014) of potato wild relatives and used environmental niche modelling (ENM) techniques to estimate potential geographic ranges of each species. Their data assigned a high priority status for collecting 32 species of wild potato, 43.8% of the species studied, due to severe gaps in the ex situ collections. Further, 20 more species were assigned medium priority for collection and only three species were determined to have good diversity representation in the ex situ collections. As part of the study, they also looked at in situ sites which were potentially threatened including that of *S. rhomboideilanceolatum* whose native habitat in Peru is increasingly threatened by road building and overgrazing. Potato wild relatives have made critical contributions to potato disease resistance, enhanced yield, and improved quality in past 50 years (Jansky et al. 2013). Therefore, it should be the obligation of present day potato researchers and breeders to do all we can to leave future potato scientists the same opportunities to explore, use, and reap benefits from the diversity of wild potato that we have today.

The geographic range of wild potato species is broadly spread out in 16 countries confined to the Americas (Mexico, United States, Costa Rica, Guatemala, Honduras, Panama, Argentina, Bolivia, Brazil, Chile, Colombia, Ecuador, Paraguay, Peru, Uruguay, and Venezuela. For thousands of years, the ecological niches of the Andes have been the natural habitat and the center of the genetic diversity of wild potato species and innumerable native varieties. Long evolutionary processes have allowed the accumulation of genetic components that are valuable resources for the improvement of potato cultivation, such as the ability to survive adverse biotic conditions (diseases, pests) and abiotic (drought, frost, climate change), as well as, vast morphological differences. However, this valuable genetic diversity is threatened by "genetic erosion." The main threats are urbanization, alternative land use, and the destruction of natural areas. Examples include overgrazing, the construction of new routes and villages, and the destruction of forests.

The domestication of potato is thought to have occurred ~10,000 years ago in the Lake Titicaca Basin on the border between Peru and Bolivia, most likely from the *S. brevicaule* complex with *S. candolleanum* as a potential ancestor (Spooner et al. 2014). The use of these wild potato species for food is likely, yet the high glycoalkaloid content needed to be dealt with in some fashion to avoid toxicity. Rumold and Aldenderfer (2016) studied early potato starch grains from tools found at a village site in the Titicaca basin (1700–3500-year-old site) which resembled freeze-dried samples rather than starch from fresh tubers. They hypothesize that a process such as drying and grinding may have played a role in the detoxification of the wild potato tubers, as in other root crops. This theory is consistent with the natural freeze drying of tubers in the Andes today known as "chuño." Recent supporting evidence for use of wild potatoes for human consumption comes from *S. jamesii* starch grains found among other 6000-year-old early human ruins in the Southwest US (Louderback and Pavlik 2017). It is interesting to note that *S. jamesii* was still used until recently as a food source for native communities in the area and that existing patterns of present day occurrence of this species suggests that early humans may

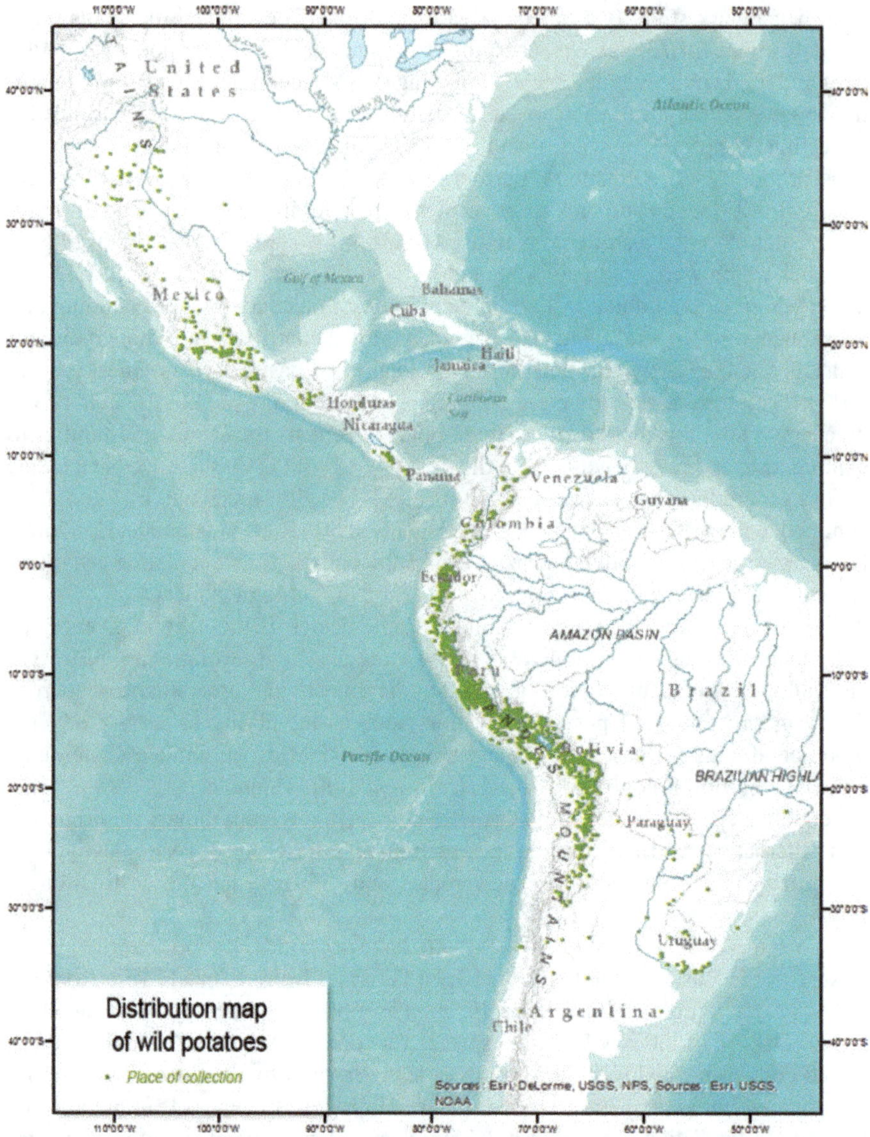

Fig. A map of the distribution of wild potato accessions in the CIP genebank. Greens dot indicates origin of the accession based on passport data (https://cipotato.org/genebankcip/process/potato/potato-wild/)

have cultivated the tubers (Kinder et al. 2017). Domestication is believed to have given rise to increased tuber size through selection and cultivation which was associated with an increase in leaf carbon fixation and transport, reduction in glycoalkaloids, adaptation to long-day photoperiod, and reduced sexual fertility (Ryoko 2015).

Fig. Examples of the diversity found in leaf morphology, shape, and color of various wild potato species

With domestication of early edible potato tubers, it is logical this also was the start not only of selection and cultivation but also of movement of selected potato tubers from one site to another. As early as 5000 years ago, maize, originating in Mexico, was already in S. America (Grobman et al. 2012), hence there is no reason to think that the north to south movement of food materials was not also reciprocated with a south to north movement of food such as cultivated potato. The movement of domesticated varieties most certainly also gave rise to the movement of new genetic material through the zone of native wild populations followed by intercrossing leading to increased diversity in the domesticated varieties and further selection for adaptive traits. These same adaptive traits and the diversity associated with them contributed to the adaptations which were selected for by early human populations and these same traits still exist in the wild populations today and will continue to be of critical value to breeders as long as potato improvement programs exist.

Fig. Examples of the diversity found in fruit morphology, shape, and color in wild potato species

Thus collecting, movement, and exchange of potato and potato wild relatives has occurred for millennia, yet was mainly confined to the Americans until the Spanish conquest when potato was collected and shipped to Europe where it underwent further selection. Wild relatives were likely also collected before the last century, but reliable documentation of the collection of wild potato for genebanking is more recent. Darwin, in his famous voyage of the Beagle, collected potato from the Chilóe Islands in Chile in the 1830s which he named *S. maglia*, now known as S. *tuberosum* (Ristaino and Pfister 2016). At this time, Darwin erroneously assumed potatoes originated from these islands due to the diversity and wide spread presence. In his journals, he referenced that the crew "*stocked up on game and potatoes.*" Fuentes (2014) references collection expeditions as yearly as the 1830–1833 by A. D'Orbigny from France in Bolivia. Perhaps the best documented collections are those done by the Vavilov Institute (Supreme Council of People's Economy) in Russia starting in 1925 by S.M. Bukasov and S.V. Juzepzuk (Loskutov 1999), after whom the wild potato species *S. bukasovii* and *S. xjuzepczukii* were named, respectively. Although detailed notes from these expeditions were never published, it is known that many species of wild potato were collected in Peru, Bolivia, and Chile (Ochoa 2004), and that some of this material still is maintained at the Vavilov Institute (Prof. N. Dzyubenko, personal communication).

Collections of wild potato continued by numerous collectors from Europe, Asia, North America, and Latin American institutions in an ad hoc manner until the early 1970s (Correll 1962; Hawkes 1990; Ochoa 1962) when CIP organized a meeting of experts to form and develop an International Germplasm Bank at CIP in Lima, Peru. Three Planning Conferences were held dealing with topics ranging from exploration, collection, conservation, and evaluation of potato genetic resources ending with the recommendation of collection strategies and to make the collected material available to researchers for their use for the benefit of humanity (International Potato Center 1973–1974, 1976, 1979). In the last 50 years, CIP has organized or been involved in ~300 systematic exploration missions for potato. These included the collection of wild and native potato germplasm varieties under the direction of C.M. Ochoa and collaborating with some 245 national and international scientists including A. David, J.B. Bamberg., J.G. Hawkes, A.M. Van Harten, J.P. Hjerting, W. Hondelmann, R. Hoopes, K.A. Okada, A. Salas, D. Spooner, and J.J.C. Van Soest. These collecting missions organized from Peru, where the greatest genetic diversity of the *Solanum* section Petota (Solanaceae) wild and cultivated exists, included exploring and collecting in the 16 countries of the American continent. Numerous accessions of each of the wild potato species were collected in the unique ecological niches along the geographical distribution of each species. Of interest was the rediscovery of wild species previously described, but known only as herbarium specimens in addition to the discovery of about 20 new species (including, *S. amayanum* (Ochoa 1989a, b, c), *S. bill-hookeri* (Ochoa 1988a, b), *S. bombycinum* (Ochoa 1983a, b), *S. chilliasense* (Ochoa 1981a, b, c, d), *S. incasicum* (Ochoa 1981a, b, c, d), *S. irosinum* (Ochoa 1981a, b, c, d), *S. longiusculus* (Ochoa 1987), *S. neovavilovii* (Ochoa 1983a, b), *S. orocense* (Ochoa 1980a, b, c), *S. ortegae* (Ochoa 1998), *S. peloquinianum* (Ochoa 1980a, b, c), *S. salasianum* (Ochoa 1989a, b, c), *S. sarasarae* (Ochoa 1988a, b), *S. simplicissimum* (Ochoa 1989a, b, c), *S. sucubunense* (Ochoa 1980a, b, c), *S. tapojense* (Ochoa 1980a, b, c), *S. taulisense* (Ochoa 1981a, b, c, d)) which were characterized and described by Ochoa (Hawkes 1989, 1990; Ochoa 1962, 1990, 1999; Spooner et al. 1999, 2001). These accessions are all documented with full passport data (country and location of collection) and many are available for use globally through the CIP website (https://cipotato.org/genebank-cip/process/potato/potato-wild/).

Most collection missions only have resources to focus collections from a single site and rarely can make multiple visits to the same site. Therefore, collectors are at the mercy of annual changes in rainfall and other abiotic and biotic factors that affect flowering or tuber germination. Exceptions are the collection trips by Bamberg and del Rio (2011) and Bamberg et al. (2003) where annual expeditions have been made to monitor in situ sites of *S. jamesii* and *S. stoloniferum* (formerly *S. fendleri*) since 1992 with over 200 visits. The expeditions were initiated to recollect from sites originally sampled in 1958 and 1978 to evaluate and characterize what changes could be measured over a 40+ year period and to maximize the ex situ collection of diversity found in this species (del Rio et al. 2001). Their research has yielded insight into the collection from wild potato, and perhaps, upset some long-held dogma. For example, as expected, they found more diversity using random amplified polymorphic DNAs (RAPDs) in seed collections, compared with the collection of tubers from individual plants in *S. jamesii*, but the opposite phenomena was observed for *S. stoloniferum*, where collection of tubers had greater diversity than seeds (Bamberg et al. 2009). In another report (del Rio et al. 1997), they found using RAPD markers, there were significant differences in allele frequencies between genebank-conserved material and recollected material from the same site in all seven comparisons of *S. jamesii* (diploid outcrosser), and 12 of 16 comparisons within *S. stoloniferum* (tetraploid inbreeder). While in situ biologists would conclude that this makes sense, since the argument is that evolutionary processes continue in situ, for the authors, this meant that recollecting from the same site was not only a good idea but also is desirable to capture new alleles (Bamberg et al. 2003). Other observations included a comparison of "easy" (road side collection sites) versus "remote" (sites one needed to hike to) where both collection methods yielded valuable alleles (Bamberg et al. 2010) and that patterns of geographic structure were not valuable predictors of genetic differences in these two species inferring that wide sampling of each population was needed to capture diversity.

International Fora for Plant Genetic Resources

The Convention of Biological Diversity (https://www.cbd.int/; CBD) is the most significant international treaty for the access to genetic resources globally; yet, it has had a negative impact on the collection of potato germplasm throughout the native range of wild potato. This is due to CBD reaffirming the sovereign rights of a nation to the genetic resources falling within the country boundaries, and hence, the country of origin has exclusive rights over terms for accessing genetic resources, including cultivated and wild relatives of potato, found within its boundaries. The treaty also reaffirmed the country of origin's right to "*fair and equitable sharing of benefits arising*" from the use of their genetic resources. CBD further asserts that nations are responsible for the long-term protection and *access* to genetic resources over which they have jurisdiction. CBD also acknowledged that access to, and sharing of, genetic resources, and *relevant technologies*, are critical for meeting food and health needs of a growing world population. Thus, it is a country's responsibility to not only protect but also make available for use and share in the benefits from such use of the genetic resources it holds. Unfortunately, most countries, and the international community generally, have been ill prepared for the responsibility of regulating access and for ensuring the sharing of benefits by third-party use of genetic resources, such as germplasm related to cultivated and wild potato relatives. Mechanisms to grant permission for the collection of plant genetic resources have taken time to implement, and in the meantime the collecting of plant genetic resources ceased. The challenge has been how to make genetic resources available for use on the one hand, while on the other, to ensure equitable sharing of benefits from the commercial utilization of the genetic resources, or derivatives thereof, by multiple third-party stakeholders' years later. Further, definition of the terms under which genetic resources could be accessed and benefits shared have been defined in the Nagoya Protocol on Access and Benefit Sharing (https://www.cbd.int/abs/, *Nagoya*). All potato species, except for *S. phureja*, are covered by the International Treaty for Plant Genetic Resources for Food and Agriculture (https://www.fao.org/plant-treaty/; ITPGRFA) which provides a legal framework to distribute the genetic material for research, training, and breeding with the acceptance of a Standard Material Transfer Agreement (SMTA) by the recipient.

Presently, there is a lack of a defined, functional mechanism for granting permission for collecting germplasm from the wild in most Latin American nations. The collection of potato germplasm for ex situ conservation in the countries where the clear majority of cultivated potato landraces and wild relatives originated and exist, has been virtually nonexistent since 2000. The exception is the United States of America (USA), which is not a party to CBD, and hence germplasm from potato wild relatives have continued to be collected in the USA (Bamberg et al. 2003), although only two species [*S. stoloniferum* (previously *S. fendleri*) and *S. jamesii*] are confirmed to exist within the boundaries of the USA. In 2016, a significant change occurred when the Peruvian National Institute for Innovative Agriculture (Instituto Nacional de Innovacion Agraria, INIA) was granted a permit for the collection of potato wild relatives within Peru, and in 2017 three collection missions were carried out collecting more than 70 populations of 14+ species (putatively *S. acaule, S. acroscopicum, S. aymaraesense, S. bukasovii, S. coelestispetalum, S. marinasense, S. pillahuatense S. raphanifolium, S. sandemani, S. sparsipilum, S. tacnaense, S. urubambae, S. velardei, S. yungasense*) under the auspices of the ITPGRFA. It is hoped that this will provide a model for further collections of wild and cultivated potato within its native range.

The collection of potato in its native range should be facilitated by the inclusion of cultivated potato (except for *S. phureja*) and all wild potato species in the list of Annex 1 crops under the ITPGRFA. The ITPGRFA established a global Multilateral System (MLS) that clearly defines the terms under which access for the use of genetic resources and their derivatives would be granted for use in research, training, and breeding under a Standard Material Transfer Agreement (SMTA) (http://www.fao.org/plant-treaty/areas-of-work/the-multilateral-system/the-smta/en/), which clearly outlines benefit sharing options. Of nations where wild potato species are native, only Colombia and Mexico are not parties to ITPGRFA, and thus, it is hoped that further facilitated access to collecting, conservation, and use of cultivated and wild potato species for research, training, and breeding will commence. For Colombia and Mexico, collecting, access, and use of native potato germplasm would be subject to bilateral agreements under the Nagoya protocol framework. To our knowledge, no other collections of wild potato species for inclusion in the MLS of the ITPGRFA have been carried out since the 1990s.

Methods of Conservation of Potato Germplasm

A major challenge for ex situ genebanks is to maintain the diversity held, often thousands of different accessions, in a form where every accession is readily available and accessible for use at a moment's notice. Collections of cultivated crops are often composed of various traditional varieties, along with improved cultivars, which often do not produce large quantities of seeds/tubers making them challenging to reproduce in field conditions. Additionally, in general, wild species are more challenging for ex situ genebanks to regenerate than cultivated due to the multitude of ecological niches they have become adapted to, often requiring unique environmental conditions (i.e., photoperiods) to thrive and reproduce, along with some genotypes/species, which genetically produce limited flowers, seeds, or tubers.

With the lifespan of most research grants being less than 5 years, delays in obtaining germplasm can mean missing critical planting windows which could result in a year's delay or failure to meet deliverables for a project. Thus, availability of material from the genebanks is critical. Most crop genebanks maintain their germplasm collections as botanical seed which for the majority of plants, including potato, stores very well and can be made available with a few weeks' notice if adequate seeds exist for that accession. Genebanks will only provide seed to requestors if they have adequate seed in cold storage with good viability to ensure that the material is safeguarded and there is enough seed to keep regenerating the accession. However, unlike most crops, cultivated potato is also often maintained as clones to preserve the unique allelic combinations in the selected material which poses additional challenges for maintenance and distribution.

Where desirable allelic combinations have not been fixed by selection, as in wild potato germplasm, the collections are maintained as botanical seed or true-potato-seed (TPS). A representative number of individuals (20–50) are typically collected from a wild population and seed is regenerated and combined to form a unique genebank accession of heterogeneous seed theoretically representative of the alleles found in that population. The accession is then distributed as a population of 50–100 seed where each seed represents a distinct genotype from that population. Thus, what is distributed is a heterogeneous mix of genotypes (seed) with each genotype representing a portion of the genetic make-up of the accession or population. Different collections of the same species (individual accessions) may or may not contain the same trait(s) desired by a breeder or even have this trait at the same frequency in different seed populations distributed. For example, when assessing late blight (*Phytophthora infestans*) resistance in wild potato accessions, Perez et al. (2001) found very high uniform resistance in three accessions of *S. orophilum* (100%, 98%, and 96% resistant individuals, respectively from 48 seedlings from each accession); whereas, in three accessions of *S. wittmachii*, resistance was high in one accession but very low in another (75%, 46% and 0.02% resistant individuals, respectively from 48 seedlings from each accession).

Unlike the wild species, cultivated potato is generally maintained as clones; however, in some cases seed populations of cultivated material may be just as useful to users. One of the main advantages of maintaining cultivated potato as seed is the overall cost and staffing required to maintain a seed collection is greatly reduced compared to in vitro collections which require subculturing every year or two. Further, sterile laboratory conditions with specialized equipment are not needed to reproduce potato if maintained as seed rather than in vitro clones. Seed populations can be produced and if adequate seed is obtained, a curator would not need to regenerate the accession for several decades unless the vigor of the seed decreases or the seeds are exhausted. Additionally, seed populations would have more diversity and new allele combinations that could lead to new trait discovery.

In contrast to maintenance of cultivated material as seed, maintenance of cultivated potato as a clone, preserving the unique allelic makeup of the landrace or variety, has advantages for either direct planting or facilitating the transfer of unique allelic combinations to breeding or research programs. Maintenance of clonal potato collections can be in the field, greenhouse, or in vitro as tissue culture. If maintained in the field or greenhouse, the collection is harvested, tubers stored and annually replanted as maintaining tuber viability for longer than a year is unreliable. Field or greenhouse maintenance of the collection also carries a risk of losing accessions due to biotic (insects and diseases) or abiotic (hail or wind) challenges. Therefore, many clonal potato collections are maintained in tissue culture as in vitro plantlets, which virtually eliminates the challenges of external biotic and abiotic factors. In vitro maintenance also facilitates international distribution of the collection as tissue culture plants are maintained in a sterile environment, and once cleaned of diseases, they can be certified as disease-free material for import purposes internationally.

The cultivated potato germplasm collection at CIP originated in the early 1970s with material collected or donated from around the world. In the early 1980s, research was initiated to place the collection into in vitro culture. Few in vitro collections have been maintained for 30 years or that have been used as extensively as the CIP cultivated potato collection. CIP currently maintains the largest in vitro potato collection with 8354 potato in vitro accessions (as of October 2018), the majority of which (89.8%) are landraces (native potatoes or "papa nativas" in Spanish) originating mainly from the Andean region, with the remaining accessions being improved varieties and breeding lines.

Since the vast majority of potato germplasm in genebanks around the world was collected over 20 years ago, and some, such as the collection at the Vavilov Institute in Russia, almost 90 years ago, most potato ex situ collections would be virtually impossible to replicate or replace today if lost. For this reason, Standards 4.9 of the Genebank Standards for Plant Genetic Resources for Food and Agriculture (FAO 2014) clearly state that "*A safety duplicate sample for every original accession should be stored in a geographically distant area, under the same or better conditions than those in the original genebank.*" CIP, along with other genebanks in the CGIAR, whose collections are held in trust under the ITPGRFA, takes this a step further with a guideline for backing up the collection in two distinct locations, one nationally and a second backup internationally. In the case of TPS accessions, such as the wild relatives of potato, a sealed package of seed equivalent to one to three times what is needed for a regeneration are backed up in a remote island archipelago in Norway at the Svalbard Global Seed Vault (https://www.nordgen.org/sgsv/, SGSV). The SGSV was established 10 years ago and funded by the Norwegian government with operations coordinated by the Nordic Gene Resource Centre and the Global Crop Diversity Trust. Shipment and coordination of all seed deposits are done by the Nordic Gene Resource Centre under a MTA which clearly states that the objective of the vault is "*...to provide a safety net for the international conservation system of plant genetic resources, and to contribute to securing the maximum amount of plant genetic diversity of importance to humanity for the long-term....*" (https://www.nordgen.org/sgsv/index.php?page=welcome). Under this agreement, the material is kept by the SGSV as a *black-box* safety back-up which denotes that the depositor retains full ownership and rights to the material, with the back-up facility holding the material only as a service of storage for the depositor. As of the end of 2017, SGSV contained almost 900,000 samples of seed from different crops including seed from 136 species of wild potato, five wild potato natural hybrids and six species of cultivated potato (taxonomy according to Hawkes 1990). Under conditions at the SGSV (airtight containers, properly dried seed, stored at −18 °C), good quality TPS can survive and maintain viability for 50–100 years.

Unfortunately, there are no analogous safety backup facilities to the SGSV for clonal material. In the case of clonal collections maintained in the field, the expense is often too high to allow distinct duplicate field planting sites, and hence, these collections are all too often not backed up and diminish over time. In the case of in vitro collections, the collections can be safely backed-up by shipping duplicate copies of in vitro cultures to a distant location for safety back-up. The facility backing-up the in vitro material generally needs to have lighted controlled temperature chambers (5–10 °C) for sustained slow growth of the material. In the case of the back-up for the USDA National Potato Genebank in vitro collection, cultures were shipped to the National Laboratory for Genetic Resources Preservation in Fort Collins, CO, where they are maintained as mini-tubers at 5 °C requiring subculture and reinitiating of the mini-tubers once every 2–3 years. For other in vitro potato collections, such as the one at CIP, the entire collection is shipped periodically (semi-annually) to two sites, one in a geographically distinct location in the high Andean Plateau at a CIP field station in Huancayo, Peru and the other facility outside of Peru at the *Brazilian Agricultural Research Corporation* (EMBRAPA) labs in Brasilia, Brazil. At both sites, the material is kept in slow growth conditions at 7 °C where the cultures can survive approximately 2 years without subculturing. Again, as with the seed deposits at the SGSV, these materials are stored in a black-box arrangement.

As genebanks are the stewards of the genetic resources that they maintain, with this stewardship comes great responsibility to ensure that no known pathogens are introduced into other countries with the distribution of germplasm creating a new disease problem. Vegetatively propagated crops such as potato, sugarcane, cassava, sweetpotato, yam, banana, citrus, and strawberry can be challenging because they must be multiplied continuously when maintained in the field to keep them in optimal growth. Although field maintenance of any crop is subject to abiotic challenges, as well as, biotic challenges such as diseases and pests, and with each multiplication or movement from one field to another, there are ample opportunities for the introduction of viruses and viroids (Sastry and Zitter 2014). Diseases caused by plant viruses limit sustainable production of vegetatively propagated staple food crops and in developing countries, this problem is confounded as agronomic practices to limit disease incidence and spread is marginal if done at all. As well, the use of new clean planting material to lessen virus titers is a luxury which most small holder farmers cannot afford. Additionally, low virus titers in planting materials generally display low symptom severity which makes infected plants not easily recognized as disease containing, and subsequently problematic if used for vegetative propagation (Bosch et al. 2007) because it can further perpetuate the diseases. Often human actions and climate change can be linked to the spread and outbreak of disease in conjunction with the movement and exchange of diseased plants and the international food supply chain are drivers of new disease threats (Wilkinson et al. 2011).

Actual examples of the human spread of pathogens with potato germplasm, other than by word of mouth, are difficult to find, yet published reports do exist to verify the potential challenge. The recent introduction of East Africa cassava mosaic-like viruses from African to the South West Indian Ocean islands follows human and insect transmission of the disease (De Bruyn et al. 2012), while the spread of the disease to Oman is likely directly related to human movement of plant material in the 1960s (Khan et al. 2012). Spread of viruses across international borders can also happen by insect vectors as in the case of the expansion and subsequent pandemic of the cassava mosaic disease (CMD) in the Great Lakes region of East and Central Africa which followed tightly the outbreaks of whitefly populations (Legg et al. 2011).

Therefore, for international distribution of germplasm from genebanks, plant material is tested, treated to eliminate diseases and certified to be clean and free of pathogens of phytosanitary importance as defined by import permits prior to transporting across any international border. For virus-free production of TPS, regeneration is carried out under strict sanitary conditions in a controlled screen house with limited access and maintained free of insects and thus vectors for viruses. Arracacha virus B–oca (arracacha and oca are two Andean Root and Tubers crops cultivated and eaten throughout Peru) strain (AVB-O), potato virus T (PVT), potato spindle tuber viroid (PSTVd), and alfalfa mosaic virus (AMV) have been shown to transmit virus particles to seed when pollen from an infected plant is used for pollinations (Jones 1982; Valkonen et al. 1992; FAO/IPGRI Technical Guidelines for the Safe Movement of Germplasm n.d.)]. Therefore, at CIP, for individual TPS production mother plants are tested pre-flowering for these seed transmitted viruses (AVB-O, PVT, PSTVd, and AMV) as well as for Andean potato latent virus (APLV), potato yellowing virus (PYV), and tobacco mosaic virus (TMV). If an individual mother plant tests positive for any of the viruses listed above, it is discarded and not used as a pollen donor or as a maternal parent for TPS production.

Pathogen elimination from clonal material is much more laborious, costly, and time-consuming. This is one rationale for in vitro culture as once the plantlets are clean, the in vitro culture methodology will maintain the investment of the phytosanitary cleaning. If endogenous bacteria are present in the cultures, these are usually eliminated with antibiotics prior to the elimination of viruses. A typical and successful virus elimination scheme starts with placing the in vitro cultures into thermotherapy (32–34 °C for 1 month), followed by isolation of 0.1–0.3 mm meristems and then plantlet regeneration. The regenerated plants are tested for the presence of viruses by serological, molecular, host range testing, and grafting onto indicator plants for final assurance of virus-free status. The entire process takes 12–24 months for one round and longer if not successful on the first run. At CIP, this method is highly efficient with 90% of all potato material going through virus elimination becoming virus-free after one run. The viruses of import/export importance in clonal material include potato virus T (PVT), potato spindle tuber viroid (PSTVd), potato virus X (PVX), potato virus Y (PVY), potato leaf roll virus (PLRV), potato virus S (PVS), Andean potato mottle virus (APMV), Andean potato latent virus (APLV), potato yellowing virus (PYV), and Arracacha virus B (AVB-O).

Advances in next-generation sequencing (NGS) hold great promise for reducing the labor, time, and cost of disease screening in germplasm collections. One such technology is the detection of viruses using small RNA sequencing and assembly (sRSA) technology. This technology is being tested on potato and it is hoped to be able to reduce the time necessary for phytosanitary cleaning of clonal potato germplasm from 12–24 months to less than 6 months (Kreuze 2014). Importantly, this technology could be invaluable in the future as new viruses or diseases of import importance develop so that entire germplasm collections can be rescreened quickly and efficiently ensuring the germplasm is readily available when needed.

The maintenance of in vitro potato clonal material has its advantages, including being readily available and the maintenance of disease-free status, but it is also undisputedly one of the most expensive long-term conservation methods available. This is due to the need for sterile plant tissue culture facilities, climate-controlled growth rooms, and highly trained personnel. Maintenance of clonal material as seed *is* one alternative, yet to maintain the allelic combinations present in an accession, maintenance of the clone is the only option. Cryopreservation, freezing tissue in liquid nitrogen at −196 °C, has been used for the long-term preservation of clonal plant genetic resources. Although research in the cryopreservation of plants has been ongoing for over 50 years (Reed 2008) and high confidence exists with the technology in animal and human systems (Di Santo et al. 2012), the application of long-term preservation of plant material, despite many successes, still has its skeptics. Limited long-term studies of plant material have shown largely positive and optimistic results (Volk et al. 2008), which support the theoretical long-term cryopreservation of plant meristems for centuries. Potato was one of the early crop plants used in cryopreservation research (Bajaj 1977; Grout and Henshaw 1978; Towill 1981a, b) and there are reports of over 20 potato species having been tested for cryopreservation using several different methods (Kaczmarczyk et al. 2011; Vollmer et al. 2017). To date the most successful methods for potato cryopreservation of genebank accessions included the IPK genebank using dimethyl sulfoxide (DMSO) droplet vitrification (Kaczmarczyk et al. 2011) with over 1000 cryopreserved accessions and the CIP genebank using a Plant Vitrification Solution (PVS2) droplet vitrification (Vollmer et al. 2017) method with over 2500 accessions cryopreserved as of the end of 2017.

Although the number of accessions or samples distributed from potato ex situ collections tells us little about *how* the germplasm was used or if it was useful at all, it does provide a measure of interest in the collections. In marketing, if a product is not available and there is little information on the attributes and benefits of the product, few will buy or order it. The same is true with germplasm collections, if the germplasm is not phytosanitary cleaned, in a form which can be distributed or due to some other factor(s), such as regulatory hurdles, is unavailable, it is not accessible for use and hence will not be used. Unfortunately, the physical accessibility of potato accessions from global collections is scattered with only 11/23 (~50%) of the collections surveyed having distributed material internationally in the period from 2004 to 2006 (Global Strategy for Ex situ Conservation of Potato 2006). While this is the only statistics found on distribution from multiple global potato global genebanks and it predates the International Treaty for Plant Genetic Resources for Food and Agriculture (Carputo et al. 2013), it serves as a good example and is likely the status today that less than 50% of potato ex situ material held in genebanks globally are available today beyond the walls of the genebank or the borders of the nation in which the genebank resides. This is a reflection of the state of the art of many factors and the international regulatory environment is just one factor (reviewed above). We have also discussed the need for phytosanitary cleaning and a propagule which can be distributed as disease-free. TPS could be an asset and a valuable tool for genebanks lacking the technology for long-term maintenance in tissue culture, yet is capable of virus screening of mother plants for TPS regeneration.

Another major limitation to the use/distribution of potato germplasm is the lack of information publicly accessible about potato germplasm in the various genebanks. While Genesys (https://www.genesys-pgr.org/welcome) offers a one site shop for information on the location of 19,066 potato accessions comprising over 175 species from 21 different countries, as of the end of 2017 it does not have information on 3 (INIA, France; Vavilov, Russia; and NIAS, Japan) out of 5 of the largest potato genebanks identified in the Second Report on The State of the World's Plant Genetic Resources for Food and Agriculture (FAO 2010). This highlights the difficulty for users to find information on the accessions conserved. Even the European Union Potato Project Database (http://ecpgr.cgn.wur.nl/eupotato/) does not contain information on the French or United Kingdom potato collections, yet it has information from both CIP and USA collections. If information on the holdings of the global potato germplasm collections is not readily available, or if available but not in a universal language, it will be impossible for most users to find information on what is held in ex situ collections.

As mentioned above, one measure of the value of genetic resources conservation is whether the collection is desired by the user community. Using the *in trust* potato collection at CIP as an example of the value of conservation of diversity of ex situ collections to the breeding and research community, since 2007, CIP has received 1,399 requests for potato germplasm. A total of 48,533 samples were distributed in the last 10 years (2007–2017), following the legal framework of the ITPGRFA. This represents over 1,200 unique accessions per year or almost 25% of the *in-trust* collection distributed annually. Material was sent to 96 countries with Peru, USA, Australia, and China requesting the most potato germplasm. All material is transferred using the Standard Material Transfer Agreement (SMTA) under the ITPGRFA. It could be expected that in the genomics era, DNA distributions would steadily increase; however, there is no trend in this as physical genetic material remains in high demand suggesting use in breeding programs still requires plants for crossing. Interestingly, in a recent (2015) survey (Shirey, unpublished data), late blight resistance and yield were the most cited traits of interest among users.

In 1997, a program was initiated to return traditional cultivars and landraces to the native, indigenous farmers in Peru (repatriation), whose ancestors preserved the potato diversity. For centuries, these potato farming communities have nurtured and planted multiple, 20–40, varieties per family as an insurance policy to ensure food availability. The rationale was that in one year one group of landraces would produce well, while in other years a different group might flourish. By conserving and planting diversity, they sustained their livelihood. In recent times, many communities have lost important cultivars they had planted for generations, due to terrorism, extreme weather, climate change, and/or disease/pest pressures. The CIP repatriation program is an example of the broader user base of ex situ genebanks where the *in-trust* collection served as a reservoir specifically for the landraces the current generation of potato guardians of the Andes (Papa Arariwa) needed to revive their traditional farming systems. In some cases, it allowed them to plant again the landraces they remembered as a child. Thus, repatriation is helping to restore the diversity and productivity in the traditional Andean potato farming region (Aguilar 2016). Further, the repatriation of genebank materials has benefits beyond restoring diversity and the traditions that come with this diversity, the added benefit is the distribution of pathogen-free stocks, which can increase productivity in a single season by as much as 40% thereby contributing to food security and poverty reduction for these communities. In total, 89 communities in Peru have received over 6,000 samples representing more than 1250 accessions of native landraces or over 50% of Peruvian landraces held in the in-trust potato collection.

Characterizing Potato Diversity in Genebanks

Molecular markers, are considered fixed landmarks in a genome, and thus, can reveal crucial genetic variability (Semagn et al. 2006). Markers have been used in many crop plants to assess genetic diversity, determine population structure, establishing trait-marker associations, discover and track quantitative trait loci (QTLs), produce genetic linkage maps, assist in selection for traits, understand the influence of genotypes on phenotypes, and more, all to improve or understand crop plants.

Loss of plant species in the past centuries has triggered genetic resource conservation with a need to accurately identify each accession. Molecular tools provide an easy less laborious method to assign plant taxa, as well as, characterize certain traits (Arif et al. 2010). To address the challenge of defining the biodiversity of a crop such as potato, biochemical and molecular markers have been used, yet the wide-scale use for looking at entire ex situ collections has only recently become more feasible due to advances in next generation sequencing (NGS) which has made genotyping projects more automated and affordable for large sample sizes.

Examples of where molecular markers have been employed to gain insight in potato are included below. Markers such as isozymes, simple sequence repeats (SSRs), and rapid amplified polymorphic DNA (RAPDs) have been employed in potato for varietal identification and genetic diversity assessments (Anoumaa et al. 2017; Carputo et al. 2013; Collares et al. 2004; Ghislain et al. 2009; Hoque et al. 2013; Rocha et al. 2010; Salimi et al. 2016; Xiaoyan et al. 2016). SSR markers have also been used to support a reevaluation of the taxonomic classification and structure of the gene pool of cultivated potato into four species (Spooner et al. 2007). Amplified fragment length polymorphisms (AFLPs) were utilized to hypothesize the domestication origin for cultivated potato from the northern species of the *S. brevicaule* group (Spooner et al. 2005) which was contrary to previous reports.

Even though different types of molecular markers exist and have been employed since the 1980s, single nucleotide polymorphisms (SNPs) are increasingly used predominantly due to recent advances in genome sequencing technology, the abundance of SNPs in most crop plants, reduced labor required to collect the data, and price per data point. The affordable cost and high-throughput nature of SNP markers have made them powerful tools for genetic analysis of plant species such as potato and highly useful in breeding (Bertioli et al. 2014). Discovery of SNPs in simple genomes is relatively easy requiring collection and evaluation of sequence data; however, in complex genomes such as potato, SNP detection is more challenging due to repetitive segments of the genome and multiple ploidy levels (Mammadov et al. 2012). Genome complexity reduction methods, such as genotyping by sequencing (GBS), diversity arrays technology (DArTseq), restriction site-associated DNA sequencing (RADseq), have been developed to aid in the discovery of novel SNPs; nevertheless, it is often challenging to identify SNP markers in polyploids such as potato, cotton, canola, and wheat (Bertioli et al. 2014; Logan-Young et al. 2015; Mammadov et al. 2012) due to separating allelic versus homoeologous SNPs or determining dosage in autopolyploids, both of which increase the rate of false positives (Clevenger and Ozias-Akin 2015).

SNP arrays have been developed which allow thousands to one million genome-wide SNP markers to be assessed simultaneously in an individual assay (LaFramboise 2009), thereby reducing the cost per marker data point. The Infinium 12K V2 Potato Array contains 12,720 SNPs, including the SNPs from the original SolCAP Infinium V1 8303 Potato Array with additional markers derived from the Infinium High Confidence SNPs (69K, Hamilton et al. 2011), which were selected for improved genome coverage, candidate genes, and regions with resistance genes. Both potato SNP arrays have been used in numerous studies as a genomic tool to improve cultivated potato or gain insight on genetic attributes. The SolCAP Infinium 8303 Potato Array were selected from 69,011 high-quality SNPs derived from six commercial potato cultivars "Atlantic," "Premier Russet," "Snowden," "Bintje," "Kennebec," and "Shepody" (Hamilton et al. 2011). These SNP markers were used to measure linkage disequilibrium for genome-wide association (GWA) mapping and population structure in European diploid and tetraploid germplasm (Stich et al. 2013). Genotyping a diversity panel of 250 lines of wild species, genetic stocks, and cultivated potato revealed that changes in heterozygosity and allele dosage has not occurred in over 150 years of breeding, but clear selection for alleles in biosynthetic pathways has occurred (Hirsch et al. 2013). The SolCAP Infinium 8303 Potato Array was used to develop linkage maps (Felcher et al. 2012), genotype populations for QTL analysis (Douches et al. 2014), and assess variation in glycoalkaloid biosynthesis (Manrique-Carpintero et al. 2013, 2014). It was also used to evaluate the genetic diversity and population structure of *S. tuberosum* sbsp. *andigenum* and *S. phureja* accessions from Colombia along with identifying 23 markers associated with nine morphological traits (Berdugo-Cely et al. 2017). In a wild species study, relationships deduced from the SNP markers were generally complementary to existing taxonomic classifications for 74 *Solanum* lines representing 25 wild taxa and were also effective in resolving complex taxa boundaries among germplasm with close genetic relationships (Hardigan et al. 2015).

Sequencing efforts including expressed sequence tags (ESTs) and the reference genome for potato (Potato Genome Sequencing Consortium 2011) allowed the development of the Infinium SOLCAP array that has provided the underpinning and a valuable tool for looking more closely at potato diversity. Additionally, the ease of data collection utilizing a SNP array has allowed genotyping of large sample numbers to finally be a reality. The CIP genebank has begun a project to genotype the entire landrace in vitro potato collection utilizing the Infinium V2 12K SNP array (http://solcap.msu.edu/potato_infinium.shtml) along with low density genotyping by sequencing (GBS) of 600 wild potato accessions and over 450 cultivated potatoes. This data is forthcoming; however, it will provide unique fingerprints of a large portion of the collection for users, diversity assessment in wild and cultivated material, identification of potential duplicated material, and genetic tools for genome-wide association studies (GWAS).

An initial assessment of the Infinium V2 SNP array was made by genotyping 250 potato accessions representing 7 taxa (Ellis et al. 2018). The genome-wide SNPs on the array were well distributed across the 12 potato chromosomes ranging from 798 to 1647 SNPs per chromosome. Of the 12,720 SNPs included in the Infinium 12K V2 Potato Array, the majority yielded good-quality signal intensities that were converted into genotypes. SNPs that did not produce a signal in ≥10% of the individuals or could not be clustered were filtered out of the dataset along with SNPs noted in previous studies to be poor or questionable (http://solcap.msu.edu/potato_infinium.shtml). The three-cluster diploid calling yielded 77% (9800) of the total SNPs on the array for use in subsequent analysis. In this data set only, 2.7% of the SNPs were monomorphic with only one of the three diploid genotypic classes scored among all the accessions of the diverse taxa used in this study. Aside from these few, most of the SNPs (97.3%) were polymorphic. A previous study with the 8303-array showed similar rates with a polymorphic rate of 75% (Stich et al. 2013); yet, this study included less taxa with only 44 genotypes. As expected, the five-cluster tetraploid calling had a lower rate of monomorphic markers with only 31 of 4859 (0.6%) of the SNPs being monomorphic; however only 39.5% of the total SNPs on the array could be utilized because of the difficulty in clustering/scoring the three heterozygous classes. Heterozygosity in this data set was high in many species/genotypes and ranged from 0.1 to 81.2% (Ellis et al. 2018).

A distance matrix was calculated from the SNP data and a phylogeny was constructed along with estimates on population structure (Ellis et al. 2018). Most of the taxa did not form distinct monophyletic clades and there were mixed taxa in each clade suggesting significant gene flow, low representation of certain taxa, and/or misclassified accessions. Further, branch lengths among individuals were short suggesting low diversity between taxa and the possibility of hybrids. *S. xjuzpeczukii* (2n = 3x = 36) was the only taxa to form a distinct monophyletic clade. The accessions of *S. xjuzpeczukii* were genetically redundant. This is interesting as these accessions originated from three countries (Bolivia, Peru, and Argentina); therefore, one would expect some genetic divergence. However, given that triploids are sterile and thus represent an evolutionary dead-end they were likely spread by human migration as clones. More of the genome needs to be evaluated (i.e., reduced representation sequencing) in these triploid accessions to elucidate if genetic differences among these accessions truly exists or if they are genetic duplicates before an archival process begins of putatively redundant material.

The phylogeny further demonstrated that *S.* x*chaucha*, *S. stenotomum* subsp. *goniocalyx*, and *S. stenotomum* subsp. *stenotomum* appear to be sister taxa with ancestry to *S. phureja.* Similar suggestions have emerged in previous studies using microsatellite data (Gomez, personal communication) and are supported by the revised taxonomy of Spooner et al. (2014) where these four taxa are lumped into a single species *S. tuberosum*. The phylogenetic tree also reveals putative discrepancies in species designation of accessions which will need to be verified in the future with phenotyping data. For example, multiple *S. stenotomum* subsp. *stenotomum* (17) and *S.* x*chaucha* (9) accessions grouped with the *S. tuberosum* subsp. *andigenum* accessions. In contrast, most of the *S. stenotomum* subsp. *goniocalyx* accessions clustered together or within the *S. stenotomum* subsp. *stenotomum* group (Ellis et al. 2018).

The analysis of the SNP data on 250 accessions from the genebank provided valuable information on intra- and interspecific relationships among taxa and provided additional support for targeting the collection of phenotypic data of suspected misclassified accessions. The data suggests that some of the accessions are hybrids and that gene flow has occurred between many of the taxa. The SNP array produced unique fingerprints with a few exceptions among this diverse panel of cultivated potato. These fingerprints provide a legacy for quality management systems (QMS) of the potato collection for years into the future to assure that as accessions are handled errors do not occur, and if they do occur, fingerprints can be used to validate genetic identity. Further, these data can be used to compare accessions between genebanks to determine where unique genetic material exists on a global scale and ensure it is safely preserved for the future. Overall, the genetic analysis facilitated a better understanding of the genetic diversity, population structure, and genetic relatedness of these potato taxa.

Climate Change and Genetic Resources Collections

The increased variability of the global climate, resulting in greater extremes of temperatures punctuated by increased dry, wet, cold, warm, and generally unpredictable climatic periods, will continue to have severe impacts on crop productivity and sustainability. Estimates of the extent of temperature increases range between 2 and 6 °C (Jarvis et al. 2011). The warming climates will be accompanied by increased atmospheric CO_2, and although both factors will favor plant productivity of C_3 plants, such as potatoes, this increased productivity will be accompanied by large increases in biotic (diseases and pests) and abiotic (drought, heat, adverse and unseasonable weather) challenges that could decrease plant yield and increase crop uncertainty. Previously adapted varieties may no longer be robust in the same geographic locality due to a lack of resistance/tolerance to these biotic and abiotic factors. These factors are interrelated, stressed resistance varieties may have attenuated resistance in different environments, resistance may be needed at different developmental stages than previously required, resistance could be overwhelmed by population sizes of pests and diseases, and new pests and diseases may appear. It must also be recognized that one cannot look at a single factor such as how climate affects insect vectors and thus the spread of diseases as an effect on insects will also affect insect pollinators reducing subsequent plant populations, insect predators, and insect populations by an extended growing season. It has been estimated that with a 2 °C rise in temperature, insects will go through one to five additional life cycles/year. In general, bacteria will respond better to moisture fluctuations regardless of temperature while fungi will respond to moisture fluctuations more in cooler weather (Beed 2011).

What is needed now and in the future is genes which can be introgressed into potato to help maintain productivity with the changing climate. Crop wild relatives conserved in genebanks offer "an enormous and unimaginable potential" for the discovery of valuable and desired traits (Machida-Hirano 2015). Wild relatives of potato have been a source of disease resistance for breeding programs for over 100 years (Hawkes 1958). An example is *Phytophthora infestans* (causal agent of late blight), which is the causal agent responsible for the devastation it inflicted during the Irish potato blight in 1845–1846. It is most severe in periods of high moisture with temperatures between 7.2 and 26.8 °C. It is predicted that for each degree of warming, late blight can occur 4–7 days earlier, extending the period of infections to between 10 and 20 days. With warmer temperatures starting earlier, there is a potential threat that late blight infestation will come earlier and earlier and last longer into the season resulting in increased fungicide applications leading to a greater pressure of the evolution of resistance to the fungicides (Beed 2011). Eleven hypersensitive-type resistance genes (R genes) have been characterized from *S. demissum*, a wild relative of potato of Mexican origin, thus *S. demissum* has been used extensively as a source for late blight resistance in European and North American cultivars (Asano and Tamiya 2016; Love 1999). Other sources of resistance in wild relatives of potato include *S. bulbocastrum, S. candolleanum, S. chacoense, S. pinnatisectum*, and *S. stoloniferon* (Machida-Hirano 2015).

Warmer temperatures can affect resistance genes rendering them ineffective. In tomato, increased temperatures caused a complete breakdown of resistance to *Ralsonia solanacearum* (bacterial wilt), another severe disease in potato (Kuun et al. 2001; Tung et al. 1992). Thus, increased temperatures affect not only the pathogenic organisms and vectors but also the resistance genes. Therefore, multiple forms of resistance for breeding programs may be needed and genebanks could provide such multiple forms of resistance. Examples of potato wild relatives where bacterial wilt resistance has been found are *S. chacoense, S. commersonii, S. phureja*, and *S. stonotonum* (Asano and Tamiya 2016; Machida-Hirano 2015).

Increased temperatures will also favor an increase in pests including aphids, weevils, and potato tuber moth (Asano and Tamiya 2016), and it has been stated that temperature is likely the single most important factor in cold blooded insect behavior, distribution, development, survival, and reproduction (Bale et al. 2002). Pest resistance in potato wild relatives is associated with high levels of glycoalkaloids, dense hairs, and trichomes found in species such as *S. chacoense, S. polyadenium*, and *S. tarijense* (Jansky et al. 2009). Nematode populations can also be influenced by climate as observed with the Southern root-knot nematode (*Meloidogyne incognita*) in coffee which has expanded its range in recent years associated with changing climatic conditions (Beed 2011). Nematode resistance in potato wild relatives can be found in *S. hougasii* for Columbia root-knot nematode (Brown et al. 1991) and *S. vernei* and *S. acuale* for cyst nematode (Hawkes 1994).

In addition to biotic challenges, climatic changes also can greatly affect potato survival and yield simply by the stresses imposed independent of effects on biotic factors. Increases in early and late season frosts, extreme heat, delay or unseasonable rains and other severe weather events. Tolerance to these abiotic factors is as important as resistance to pests and diseases. *S. acuale*, a potato wild relative can form a carpet of tiny plants at 4500 m.a.s.l displaying extreme frost and drought tolerance (personal observation). Other frost tolerant species include *S. demissum* and *S. xjuzcepczukii*. Drought tolerance can be found in *S. chillonanum, S. jamesii*, and *S. okadae* (Watanabe et al. 2011) and *S. xcurtilobum* (Cabello et al. 2012).

Concluding Comments

It is fortunate that the vast majority of the diversity of cultivated potato, as well as much of that of its wild relatives, is already safely conserved in numerous genebanks around the world. These resources represent an investment to ensure potato's role in food security well into the future. These conserved potato genetic resources will provide the traits and alleles needed to continue potato's dominance on plates around the world. Increased knowledge of the molecular and biochemical pathways is critical to maximize not only the discovery but also the use of the needed attributes in breeding programs and food systems. The benefits of these ex situ collections will continue to be delivered as the hidden characteristics, some with unknown value today, continue to be uncovered. One cannot however sit back complacently because as long as there is diversity which is threatened, and not securely conserved, there will be valuable alleles, individuals, and populations that could be critical in the future.

References

Aguilar AM (2016) Informe final de siembra de papas nativas repatriadas en la comunidad capesina de Santa Cruz de Pichiu, Chana, Huari, Ancash, 51p

Anoumaa M, Yao NK, Kouam EB, Kanmegne G, Machuka E, Osama SK, Nzuki I, Kamga YB, Fonkou T, Omokolo DN (2017) Genetic diversity and core collection for potato (*Solanum tuberosum* L.) cultivars from Cameroon as revealed by SSR markers. Am J Potato Res 94:449–463

Arif IA, Bakir MA, Khan HA, Al Farhan AH, Al Homaidan AA, Bahkali AH, Al Sadoon M, Shobrak M (2010) A brief review of molecular techniquest to assess plant diversity. Int J Mol Sci 11:2079–2096

Asano K, Tamiya S (2016) Breeding of pest and disease resistant potato cultivars in Japan using classical and molecular approaches. Jpn Agric Res Q 50(1):1–6

Bajaj YPS (1977) Initiation of shoots and callus from potato-tuber sprouts and axillary buds frozen at −196 °C. Crop Improv 4:48–53

Bale JS, Masters GJ, Hodkinson ID, Awmack C, bezemer TM, Brown VK, Bitterfield J, Buse A, Coulson JC, Farrer J, Good JEG, Harrington R, Hartley S, Jones TM, Lindroth RL, Press MC, Symrnooudis I, Watt AD, Whittaker JB (2002) Herbivory in global climate change research: direct effects of rising temperatures on insect herbivores. Glob Chang Biol 8:1–16

Bamberg J, del Rio A (2011) Use of native potatoes for research and breeding. Hortscience 46(11):1444–1445

Bamberg J, del Rio A, Moreyra R (2009) Genetic consequences of clonal versus seed sampling in model populations of two wild potato species indigenous to the USA. Am J Potato Res 86:367–372

Bamberg J, del Rio A, Fernandez C, Salas A, Vega S, Zorrilla C, Roca W, Tay D (2010) Comparison of "remote" versus "easy" in situ collection locations for USA wild solanum (potato) germplasm. Am J Potato Res 87:277–284. https://doi.org/10.1007/s12230-010-9133-8

Bamberg JB, del Rio A, Huamán Z, Vega S, Salas A, Pavek J, Fernandez C, Spooner DM (2003) A decade of collecting and research on wild potatoes of the Southwest USA. Am J Potato Res 80:159–172

Beed F (2011) The impact of climate change on interdependence for microbial genetic resources for agriculture. In: Fujisaka S, Williams D, Halewood M (eds) The impact of climate change on countries interdependence on genetic resources for food and agriculture. Background Study Paper No. 48. FAO Commission on Genetic Resources for Food and Agriculture, Italy, pp 38–47

Berdugo-Cely J, Valbuena RI, Sanchez-Betancourt E, Barrero LS, Yockteng R (2017) Genetic diversity and association mapping in the Colombian central collection of *Solanum tuberosum*

L. Andigenum group using SNP markers. PLoS One 12(3):e0173039. https://doi.org/10.1371/journal.pone.0173039

Bertioli DJ, Ozias-Akins P, Chu Y, Dantas KM, Santos SP, Gouvea E, Guimaraes PM, Leal-Bertioli SC, Knapp SJ, Moretzsohn MC (2014) The use of SNP markers for linkage mapping in diploid and tetraploid peanuts. G3 (Bethesda) 4:89–96

Bosch FVD, Jeger MJ, Gilligan CA (2007) Disease control and its selection for damaging plant virus strains in vegetatively propagated staple food crops; a theoretical assessment. Proc R Soc [Biol] 247:11–18

Brown CR, Mojtahedi H, Santo GS (1991) Resistance to Columbia root-knot nematode in *Solanum* spp. and in hybrids of *S. hougasii* with tetraploid cultivated potato. Am Potato J 68:445–452

Cabello R, De Mendiburu F, Bonierbale M, Monneveux P, Roca W, Chujoy E (2012) Large-scale evaluation of potato improved varieties, genetic stocks and landraces for drought tolerance. Am J Potato Res 89:400–410

Carputo D, Alioto D, Aversano R, Garramone R, Miraglia V, Villano C, Frusciante L (2013) Genetic diversity among potato species as revealed by phenotypic resistances and SSR markers. Plant Genetic Resources 11(2): 131–139. https://doi.org/10.1017/S1479262112000500

Castañeda-Álvarez NP, de Haan S, Juárez H, Khoury CK, Achicanoy HA, Sosa CC, Bernau V, Salas A, Heider B, Simon R, Maxted N, Spooner D (2015) Ex situ conservation priorities for the wild relatives of potato (*Solanum* L. Section Petota). PLos One 10(14):e0122599

Clevenger JP, Ozias-Akin P (2015) SWEEP: a tool for filtering high quality SNPs in polyploid crops. G3 (Bethesda) 5:1797–1803

Collares EAS, Choer E, Pereira AS (2004) Characterization of potato genotypes using molecular markers. Pesqui Agropecu Bras 39:871–878

Correll DS (1962) The potato and its wild relatives. Contributions from the Texas Research Foundation 4, p 606. Texas Research Foundation, Renner, Texas

De Bruyn A, Villemot J, Lefeuvre P, Villar E, Hoareau M, Harimalala M, Abdoul-Karime AL, Abdou-Chakour C, Reynaud B, Harkins GW, Varsani A, Martin DP, Lett J-M (2012) East African cassava mosaic-like viruses from Africa to Indian Ocean Islands: molecular diversity, evolutionary history and geographical dissemination of a bipartite begomovirus. BMC Evol Biol 228(12):228–246

del Rio AH, Bamberg JB, Huaman Z, Salas A, Vega SE (1997) Assessing changes in the genetic diversity of potato gene banks. 2. In situ vs ex situ. Theor Appl Genet 95:199–204

del Rio AH, Bamberg JB, Huaman Z, Salas A, Vega SE (2001) Association of ecogeographical variables and RAPD marker variation in wild potato populations of the USA. Crop Sci 41:870–878

Douches D, Hirsch CN, Manrique-Carpintero NC, Massa AN, Coombs J, Hardigan M, Bisognin D, De Jong W, Buell CR (2014) The contribution of the Solanaceae coordinated agricultural project to potato breeding. Potato Res 57:215–224

Ellis D, Chavez O, Coombs J, Soto J, Gomez R, Douches D, Panta A, Silvestre R, Anglin NL (2018) Genetic identity in genebanks: application of the SolCap 12K SNP array in fingerprinting the global in trust potato collection. Genome 61:523–537

FAO (2010) The second report on the state of the world's plant genetic resources for food and agriculture. FAO, Rome, 368p

FAO (ed) (2014) Genebank standards for plant genetic resources for food and agriculture, rev edn. FAO, Rome

FAO/IPGRI Technical Guidelines for the Safe Movement of Germplasm. https://cropgenebank.sgrp.cgiar.org/images/file/learning_space/potato_tech_guid_safe_move_germplasm.pdf

Felcher KJ, Coombs JJ, Massa AN, Hansey CN, Hamilton JP, Veilleux RE, Buell CR, Douches DS (2012) Integration of two diploid potato linkage maps with the potato genome sequence. PLoS One 12:e36347

Fuentes XC (2014) Conserving the genetic diversity of Bolivian wild potatoes. PhD thesis, Wageningen University, Wageningen, NL, 229p, ISBN 978-94-6257-168-67

Ghislain M, Nunez J, Herrera MDR, Pignataro J, Guzman F, Bonierbale M, Spooner DM (2009) Robust and highly informative microsatellite based genetic identity kit for potato. Mol Breed 23:377–388

Global Strategy for Ex situ Conservation of Potato 2006. https://www.genebanks.org/wp-content/uploads/2017/01/Potato-Strategy-2007.pdf

Grobman A, Bonavia D, Dillehay TD, Pipernod DR, Iriartef J, Holst I (2012) Preceramic maize from Paredones and Huaca Prieta, Peru. Proc Natl Acad Sci U S A 109(5):1755–1759

Grout BWW, Henshaw GG (1978) Freeze preservation of potato shoot-tip cultures. Ann Bot 42:1227–1229

Hamilton JP, Hansey CN, Whitty BR, Stoffel K, Massa AN, Deynze AV, De Jong WS, Douches DS, Buell CR (2011) Single nucleotide polymorphism discovery in elite North American potato germplasm. BMC Genomics 12:302

Hardigan MA, Bamberg J, Buell CR, Douches DS (2015) Taxonomy and genetic differentiation among wild and cultivated germplasm of *Solanum* sect Petota. Plant Genome 1:16–270

Hawkes JG (1958) Significance of wild species and primitive forms for potato breeding. Euphytica 7:257–270

Hawkes JG (1989) Nomenclatural and taxonomic notes on the infrageneric taxa of the tuber-bearing solanums (Solanaceae). Taxon 39(3):489–492

Hawkes JG (1990) The potato: evolution, biodiversity and genetic resources. Belhaven Press, London, 259p

Hawkes JG (1994) Origins of cultivated potatoes and species relationships. In: Bradshaw JE, MacKay GR (eds) Potato genetics. CAB International, Wallingford, pp 3–42

Hijmans RJ, Spooner DM (2001) Geographic distribution of wild potato species. Am J Bot 88(11):2101–2112

Hirsch CN, Hirsch CD, Felcher K, Coombs J, Zarka D, Deynze AV, Jong WD, Veilleux RE, Jansky S, Bethke P, Douches DS, Buell CR (2013) Retrospective view of North American potato (*Solanum tuberosum* L.) breeding in the 20th and 21st centuries. G3 (Bethesda) 3:1003–1013. https://doi.org/10.1534/g3.113.005595.

Hoque ME, Huq H, Moon NJ (2013) Molecular diversity analysis in potato (*Solanum tuberosum* L.) through RAPD markers. SAARC J Agric 11:95–102

International Potato Center (1973–1974) Germplasm exploration and taxonomy of potatoes. In: Program planning conferences 1973–74, Lima, Perú, pp 5–48

International Potato Center (1976) Report of the planning conferences on the exploration and maintenance of germplasm resources, Lima, Peru, Marzo 15–19, 1976, 130p

International Potato Center (1979) Report of the planning conferences on the exploration, taxonomy and maintenance of potato germplasm III. CIP, Lima, Peru, 193p

Jansky SH, Simon R, Spooner DM (2009) A test of taxonomic predictability: resistance to Colorado potato beetle in wild relatives of potato. J Econ Entomol 102:442–431

Jansky SH, Dempewolf H, Camadro EL, Simon R, Zimnoch-Guzowska E, Bisognin DA, Bonierbale M (2013) A case for crop wild relative preservation and use in potato. Crop Sci 53:746–754

Jarvis A, Ramirez J, Hansen J, Leibing C (2011) Crop and forage genetic resources: international interdependence in the face of climate change. In: Fujisaka S, Williams D, Halewood M (eds) The impact of climate change on countries interdependence on genetic resources for food and agriculture. Background Study Paper No. 48. FAO Commission on Genetic Resources for Food and Agriculture, Italy, pp 7–83

Jones RAC (1982) Tests for transmission of four potato viruses through potato true seed. Ann Appl Biol 100:315–320

Kaczmarczyk A, Rokka V-M, Keller ERJ (2011) Potato shoot tip cryopreservation. A review. Potato Res 54:45–79. https://doi.org/10.1007/s11540-010-9169-7

Khan AJ, Akhtar S, Al-Matrushi AM, Briddon RW (2012) Introduction of East African cassava mosaic Zanzibar virus to Oman harks back to “Zanzibar, the capital of Oman”. Virus Genes 46(1):195–198. https://doi.org/10.1007/s11262-012-0838-2

Kinder DH, Adams KR, Wilson HJ (2017) *Solanum Jamesii*: evidence for cultivation of wild potato tubers by ancestral Puebloan groups. J Ethnobiol 37(2):218–240

Kreuze J (2014) siRNA deep sequencing and assembly: piecing together viral infections. In: Gullino ML, Bonants PJM (eds) Detection and diagnostics of plant pathogens, plant pathology in the 21st century, vol 5. Springer, Dordrecht, 200p. https://doi.org/10.1007/978-94-017-9020-8_2

Kuun GK, Okole B, Bornmann L (2001) Protection of phenypropanoid metabolism by prior heat treatment in *Lycopersicon esculentum* exposed to *Ralstonia solanacearum*. Plant Physiol Biochem 39:871–880

LaFramboise T (2009) Single nucleotide polymorphism arrays: a decade of biological, computational and technological advances. Nucleic Acids Res 37:4181–4193

Legg JP, Jeremiah SC, Obiero HM, Maruthi MN, Ndyetabula I, Okao-Okuja G, Bouwmeester H, Bigirimana S, Tata-Hangy W, Gashaka G, Mkamilo G, Alicai T, Kumar LP (2011) Comparing the regional epidemiology of the cassava mosaic and cassava brown streak virus pandemics in Africa. Virus Res 159(2):161–170

Logan-Young CJ, Yu JZ, Verma SK, Percy RG, Pepper AE (2015) SNP discovery in complex allotetraploid genomes (*Gossypium spp*., Malvaceae) using genotyping by sequencing. Appl Plant Sci 3:1400077. https://doi.org/10.1007/s12229-014-9146-y

Loskutov IG (1999) Vavilov and his institute. A history of the world collection of plant genetic resources in Russia. International Plant Genetic Resources Institute, Rome. ISBN 92-9043-412-0

Louderback LA, Pavlik BM (2017) Starch granule evidence for the earliest potato use in North America. Proc Natl Acad Sci U S A 114(29):7606–7610

Love SL (1999) Founding clones, major contributing ancestors, and exotic progenitors of prominent North American potato cultivars. Am J Potato Res 76:263–272

Machida-Hirano R (2015) Diversity in potato genetic resources. Breed Sci 65:26–40

Mammadov J, Aggarwal R, Buyyarapu R, Kumpatla S (2012) SNP markers and their impact on plant breeding. Int J Plant Genomics 2012:728398. https://doi.org/10.1155/2012/728398

Manrique-Carpintero NC, Tokuhisa JG, Ginzberg I, Holliday JA, Veilleux RE (2013) Sequence diversity in coding regions of candidate genes in the glycoalkaloid biosynthetic pathway of wild potato species. G3 (Bethesda) 3:1467–1479

Manrique-Carpintero NC, Tokuhisa JG, Ginzberg I, Veilleux RE (2014) Allelic variation in genes contributing to glycoalkaloid biosynthesis in a diploid interspecific population of potato. Theor Appl Genet 127:391–405

Ochoa CM (1962) Los Solanun Tuberiferos silvestres del Peru Secc. Tubernarium, sub-secc. Hyperbasarthrum), Lima, Peru

Ochoa CM (1980a) New taxa of Solanum from Peru and Bolivia. Phytologia 46(4):223–225

Ochoa CM (1980b) New tuber-bearing Solanum from Colombia. Phytologia 46(7):495–497. ISSN 0031-9430

Ochoa CM (1980c) Solanum peloquinianum: a new wild potato species from the Peruvian Andes. Am Potato J 57(1):33–35

Ochoa CM (1981a) [Solanum chilliasense, a new tuber bearing species of the Series Piurana]. Solanum chilliasense, nueva especie tuberifera de la serie Piurana. Lorentzia (Argentina) 4:9–11

Ochoa CM (1981b) [Solanum taulisense, a new Peruvian tuber bearing species]. Solanum taulisense, nueva especie tuberifera peruana. Lorentzia (Argentina). 4:13–15

Ochoa CM (1981c) Solanum irosinum, new Peruvian tuber bearing species resistant to Phytophthora infestans. Am Potato J 58(3):131–133. ISSN 0003-0589

Ochoa CM (1981d) Two new tuber-bearing Solanum from South America. Phytologia 48:229–232

Ochoa CM (1983a) Solanum bombycinum, a new tuber-bearing tetraploid species from Bolivia. Am Potato J 60(11):849–852. ISSN 0003-0589

Ochoa CM (1983b) Solanum neovavilovii: a new wild potato species from Bolivia. Am Potato J 60(12):919–923. ISSN 0003-0589

Ochoa CM (1987) Solanum longiusculus (Sect. petota), nova specie peruviana. Phytologia 63(5):329–330. ISSN 0031-9430

Ochoa CM (1988a) [Solanum sarasarae (Sect. Petota), a new species from Perú]. Solanum sarasarae (Sect. Petota), nova specie peruviana. Phytologia 64(4):245–246

Ochoa CM (1988b) Solanum bill-hookeri: new wild potato species from Peru. Am Potato J 65(12):737–740

Ochoa CM (1989a) [Solanum Ser. Simplicissima, new serie Tuberifera of Sect. Petota (Solanaceae)]. Solanum Ser. Simplicissima, nueva serie Tuberifera de la Sect. Petota (Solanaceae). Revista de la Academia Colombiana de Ciencias Exactas, Físicas y Naturales 17(65):321–323. ISSN 0370-3908

Ochoa CM (1989b) Solanum amayanum: a new wild Peruvian potato species. Am Potato J 66(1):1–4. ISSN 0003-0589

Ochoa CM (1989c) Solanum salasianum: new wild tuber-bearing species from Peru. Am Potato J 66(4):235–238. ISSN 0003-0589

Ochoa CM (1990) The potatoes of South America: Bolivia. Cambridge University Press, Cambridge, 512p. ISBN 0-521-38024-3

Ochoa CM (1998) Solanum ortegae: a new Peruvian species from Sect. petota. Phytologia 85(4):271–272. ISSN 0031-9430

Ochoa CM (1999) Las papas de Sudamérica: Perú (parte 1). Allen Press, Kansas, 1036p. ISBN 92-9060-197-3

Ochoa CM (2004) The potatoes of South America: Peru The wild species, part 1. International Potato Center, Peru, 1036p. ISBN 9290601981

Perez W, Salas A, Raymundo R, Huaman Z, Nelson R, Bonierbale M (2001) Evaluation of wild potato species for resistance to Late Blight. In: Scientist and farmer, partners in research for the 21st century, Program Report International Potato Center, Lima, Peru, 480p

Potato Genome Sequencing Consortium (2011) Genome sequence and analysis of the tuber crop potato. Nature 475:189–195

Reed BM (2008) Cryopreservation—practical considerations. In: Reed BM (ed) Plant cryopreservation a practical guide. Springer Science, New York, 513p

Ristaino JB, Pfister DH (2016) "What a painfully interesting Subject": Charles Darwin's studies of potato late blight. BioScience 66(12):1035–1045

Rocha EA, Paiva LV, de Carvalho HH, Guimaraes CT (2010) Molecular characterization and genetic diversity of potato cultivars using SSR and RAPD markers. Crop Breed Appl Biotechnol 10:204–210

Rumold CU, Aldenderfer MS (2016) Late archaic–early formative period microbotanical evidencefor potato at Jiskairumoko in the Titicaca Basin of southern Peru. Proc Natl Acad Sci U S A 13(48):13672–13677

Ryoko M-H (2015) Diversity in potato genetic resources. Breed Sci 65:26–40

Salimi H, bahar M, Mirlohi A, Talebi M (2016) Assessment of the genetic diversity among potato cultivars from different geographical áreas using the genomic and EST microsatellites. Iran J Biotechnol 14:270–277

Di Santo M, Tarozzi N, Nadalini N, Borini A. (2012) Human sperm cryopreservation: Update on techniques, effect on DNA integrity, and implications for ART. Adv Urol 2012:854837. https://doi.org/10.1155/2012/854837.

Sastry KS, Zitter TA (2014) Plant virus and viroid diseases in the tropics, volume 2: epidemiology and management. Springer, Dordrecht, pp 149–480. https://doi.org/10.1007/978-94-007-7820-7_2

Semagn K, Bjornstad A, Ndjiondjop MN (2006) An overview of molecular marker methods for plants. Afr J Biotechnol 5:2540–2568

Spooner DM, Salas Lopez A, Huaman Z, Hijmans RJ (1999) Wild potato collecting expedition in southern Peru (Department of Apurimac, Arequipa, Cusco, Moquegua, Puno, Tacna) in 1998: taxonomy and new genetic resources. Am J Potato Res 76(3):103–119. ISSN 1099-209X

Spooner DM, Salas AR, Huaman Z, Torres Maita RV, Hoekstra R, Schuler K, Hijmans RJ (2001) Taxonomy and new collections of wild potato species in central and southern Perú. Am J Potato Res 78(3):197–208. ISSN 1099-209X

Spooner DM, McLean K, Ramsay G, Waugh R, Bryan GJ (2005) A single domestication for potato based on multilocus amplified fragment length polymorphism genotyping. Proc Natl Acad Sci U S A 102:14694–14699

Spooner DM, Nunez J, Trujillo G, Herrera MDR, Guzman F, Ghislain M (2007) Extensive simple sequence repeat genotyping of potato landraces supports a major reevaluation of their gene pool structure and classification. Proc Natl Acad Sci U S A 104:19398–19403

Spooner DM, Ghislain M, Simon R, Jansky SH, Gavrilenko T (2014) Systematics, diversity, genetics, and evolution of wild and cultivated potatoes. Bot Rev 80:283–383. https://doi.org/10.1007/s12229-014-9146-y.

Stich B, Urbany C, Hoffmann P, Gebhardt C (2013) Population structure and linkage disequilibrium in diploid and tetraploid potato revealed by genome wide high density genotyping using the SolCap SNP array. Plant Breed 132:718–724

Towill LE (1981a) *Solanum etuberosum*—a model for studying the cryobiology of shoot-tips in the tuberbearing Solanum species. Plant Sci Lett 20:315–324

Towill LE (1981b) Survival at low temperatures of shoot-tips from cultivars of *Solanum tuberosum* group Tuberosum. CryoLett 2:373–382

Tung PX, Hermsen JG, Zaag P, Vander S, Schmiedche P (1992) Effects of heat tolerance on expression of resistance to *Pseudomonas solanacearum* in potato. Potato Res 35:321–328

Valkonen JPT, Pehu E, Watanabe K (1992) Symptom expression and seed transmission of alfalfa mosaic virus and potato yellowing virus (SB-22) in *Solanum brevidens* and *Setuberosum.* Potato Res 35:403–410

Volk GM, Wadell J, Bonnart R, Towill L, Ellis D, Lauffman M (2008) High viability of dormant *Malus* buds after 10 years of storage in liquid nitrogen vapour. CryoLetters 29:89–94

Vollmer R, Villagaray R, Egúsquiza V, Cardenas JE, Castro M, Chavez O, Anglin NA, Ellis D (2017) A large-scale viability assessment of the potato cryobank at the International Potato Center (CIP). In Vitro Cell Dev Biol Plant 53(4):309–317. https://doi.org/10.1007/s11627-017-9846-1

Watanabe KN, Kikuchi A, Shimazaki T, Asahima M (2011) Salt and drought stress tolerance in transgenic potatoes and wild species. Potato Res 54:319–324

Wilkinson K, Grant WP, Green LE, Hunter S, Jeger MJ, Lowe P, Medley GF, Mills P, Philipson J, Poppy GM, Waage J (2011) Infectious diseases of animals and plants: an interdisciplinary approach. Philos Trans R Soc B Biol Sci 366(1573):1933–1942

Xiaoyan S, Chunzhi Z, Ying L, Shuangshuang F, Qing Y, Sanwen H (2016) SSR analysis of genetic diversity among 192 diploid potato cultivars. Hortic Plant J 2:163–171

3

Fungal, Oomycete, and Plasmodiophorid Diseases of Potato

Birgit Adolf, Jorge Andrade-Piedra, Francisco Bittara Molina, Jaroslaw Przetakiewicz, Hans Hausladen, Peter Kromann, Alison Lees, Hannele Lindqvist-Kreuze, Willmer Perez, and Gary A. Secor

Abstract This chapter discusses the major potato diseases worldwide: late blight, early blight, wart, and powdery scab. **Late blight**, caused by the oomycete *Phytophthora infestans*, continues to be the main biotic constraint of potato production. Annual losses have been estimated to be about €6.1 billion, with major consequences to food security, especially in developing countries. Symptoms of the disease can be seen in leaves (water-soaked light to dark brown spots), stems (brown spots), and tubers (slightly depressed areas with reddish-brown color). High humidity and mild temperatures are essential for disease development and, under optimal conditions, the disease can destroy a field in a few days. *Phytophthora infestans* evolves continuously, mainly through recombination and migration from other

B. Adolf · H. Hausladen
Technische Universität München, Freising, Germany
e-mail: b.adolf@lrz.tum.de; h.hausladen@lrz.tum.de

J. Andrade-Piedra · H. Lindqvist-Kreuze (✉) · W. Perez
International Potato Center, Lima, Peru
e-mail: j.andrade@cgiar.org; h.lindqvist-kreuze@cgiar.org

F. Bittara Molina
J. R. Simplot Company, Plant Sciences, Boise, ID, USA
e-mail: Francisco.BittaraMolina@simplot.com

J. Przetakiewicz
Plant Breeding and Acclimatization Institute, Radzikow, Blonie, Poland
e-mail: j.przetakiewicz@ihar.edu.pl

P. Kromann
International Potato Center, Nairobi, Kenya
e-mail: p.kromann@cgiar.org

A. Lees
The James Hutton Institute, Invergowrie, Dundee, UK
e-mail: alison.lees@hutton.ac.uk

G. A. Secor
Department of Plant Pathology, North Dakota State University, Fargo, ND, USA
e-mail: gary.secor@ndsu.edu

areas. Thus, monitoring of *P. infestans* populations is critical for the design of effective management strategies. Fungicides remain as the most common tactic for late blight management, but environmental considerations are increasing the pressure to use host resistance, sanitation, and other measures. New solutions being developed to manage late blight include, among others, smart phone-based decision support systems linked to portable molecular diagnostics kits that can disseminate disease information rapidly to a large number of farmers. Emerging research topics on *P. infestans* include the role of the pathogen–microbiota interaction in promotion or suppression of the disease, as well as the metabolism of *P. infestans*.

The fungus *Alternaria solani* is the main pathogen causing **early blight** on potatoes. Early blight can be found in most potato-growing countries. Typical symptoms on the leaves are dark brown to black spots with concentric rings (target spot). In susceptible potato cultivars in particular, as well as in locations (especially in warmer areas) with increased occurrence of *A. solani*, the disease can cause considerable yield losses. Integrated pest management to control early blight requires the implementation of several approaches. The disease is primarily controlled by the use of cultural practices (to reduce the soil born inoculum), less susceptible cultivars and the use of pesticides. But there is a loss in sensitivity toward two groups of fungicides described. The loss in sensitivity towards succinate dehydrogenase inhibitor (SDHI) and Quinone outside inhibitor (QoI) fungicides are caused by different point mutations. In many countries the occurrence of SDHI and QoI mutants is reported. Therefore, the control of early blight will be a considerable challenge in the future. The increasing importance of early blight in potatoes is due to a number of factors.

Synchytrium endobioticum is a soil-borne biotrophic fungus causing **potato wart** disease of cultivated potato. The fungus originates from the Andean zones of South America, from where it spread first to Europe in the late of the nineteenth century. Presently, the geographical distribution of this pathogen includes almost all European and Mediterranean Plant Protection Organization (EPPO) countries, Asia, North and South America as well as Oceania (New Zealand). The typical symptoms of cauliflower-like galls could develop on all meristematic tissues of potato except roots. *S. endobioticum* produces summer sporangia with mobile zoospores that can move in the soil. Winter (resting) sporangia are the dormant structures by which the fungus disperses to establish new infections. They can survive more than 40 years without plant hosts. The pathogen does not produce hyphae. Its long persistence in soil and the severe losses it inflicts to potato crops have prompted its inclusion into the A2 quarantine list of EPPO. Since the discovery of pathotype 2(G1) in Germany, more than 40 pathotypes were reported in Europe. In Europe, pathotypes 1(D1), 2(G1), 6(O1), and 18(T1) are the most relevant. Other pathotypes occur mainly in the rainy mountainous areas of central and eastern Europe. *S. endobioticum* is a still serious problem for crop production in countries with moderate climates. The strategies to confine the disease are strict quarantine and phytosanitary measures, and the cultivation of resistance cultivars of potato.

Spongospora subterranea causes root galling and tuber **powdery scab** leading to quality and yield losses in seed and ware crops worldwide and is also important as the natural vector of potato mop-top virus (PMTV), an economically important tuber blemish disease of potato. *S. subterranea* spreads by movement of infected seed tubers and soil and can survive long periods in soils and some asymptomatic hosts. Powdery scab is particularly favored by cool, damp conditions and is an intractable disease. Avoidance is the best control for powdery scab, but once soil is infested with *S. subterranea*, cultural practices and chemical treatments are ineffective control methods, and host resistance appears to be the most promising mechanism for long-term management. Although sources of host resistance have been identified, they are not widely deployed in practice. *S. subterranea* is an unculturable biotroph, making research difficult. Recent progress in understanding the biology of *S. subterranea* as a result of the application of basic molecular techniques, and future opportunities to further advance knowledge of this understudied pathogen, and the virus that it vectors, are included.

Late Blight

Hannele Lindqvist-Kreuze, Willmer Perez, Peter Kromann and Jorge Andrade-Piedra

Causal Organism

Late blight is caused by *Phytophthora infestans* Mont de Bary. It was previously classified as a fungus due to the superficial resemblance to filamentous fungi but is now classified as oomycete in the kingdom of stramenopiles (Kamoun et al. 2014). The vegetative stage of the mycelium in *P. infestans* is diploid, while in true fungi it is haploid. However, recent research has shown that in the modern-day lineages the progenies from sexual *P. infestans* populations are diploid, but the most important pandemic clonal lineages are triploid (Li et al. 2017). Virulence of oomycetes depends on large, rapidly evolving protein families including extracellular toxins, hydrolytic enzymes, and cell entering effectors that help the pathogen suppress the host plant defenses and gain nutrition from the host (Jiang and Tyler 2012). Elicitins are an example of structurally conserved extracellular proteins of *P. infestans* that have a function in the sequestration of sterols from the host plant, but can also act as pathogen-associated molecular patterns (PAMPs), and as such can activate PAMP triggered immunity (PTI) (Du et al. 2015). *P. infestans* secretes large numbers of effectors: apoplastic effectors, such as EPIC1, interact with the host cell wall, host proteases and other defense-related molecules in the host extracellular space, while cytoplasmic effectors, the RxLR proteins, and CRNs (crinkling and necrosis-induced proteins) function inside the plant cells (reviewed in Whisson et al. 2016). RxLR effectors act as activators of plant immunity, resulting in effector triggered immunity (ETI) (Oh et al. 2009; Wang et al. 2017a), while the apoplastic effectors, similarly to elicitins, act as activators of the PTI (Domazakis et al. 2017). Recent research has also shown that some *P. infestans* effectors can target host proteins whose activity enhances susceptibility possibly through the inhibition of positive regulators of immunity or promote the activity of susceptibility (S), that in turn can negatively regulate immunity (Boevink et al. 2016). The effector genes locate mostly in the gene sparse regions of the genome, that are rich in repetitive sequences and are rapidly evolving, probably enabling the evolutionary arms race between *P. infestans* and the host plant (Haas et al. 2009; Dong et al. 2015).

Symptoms

The asexual, aerially dispersed sporangia are responsible for most of the devastating epidemics on potato. When the sporangia lands on a plant surface it can germinate directly or first form zoospores, which encyst, germinate, and penetrate the host tissue (reviewed by Fry et al. 2015). This stage of infection is unnoticeable to the naked eye, but inside the plant cell a repertoire of molecular interactions takes place. After penetration and adhesion, the pathogen forms haustoria inside the plant cells, from where it secretes effector proteins (reviewed in Nowicki et al. 2012; Whisson et al. 2016; Wang et al. 2017a). At this biotrophic stage *P. infestans* requires living cells to obtain nutrients.

The first visible symptoms appear within 2–3 days when the pathogen switches to the necrotrophic stage. In leaves, lesions are light to dark brown in color, water-soaked, irregularly shaped, sometimes surrounded by a yellow halo and not limited by leaf veins. Symptoms typically begin to develop where water accumulates near the leaf edges or tips and in stems near petioles. Affected tubers show irregular, slightly depressed areas with brown color. In a cross-section, finger-like extensions can be seen from the external surface to the tuber medulla (Perez and Forbes 2010).

Fig. Lemon-shaped sporangium of *Phytophthora infestans*

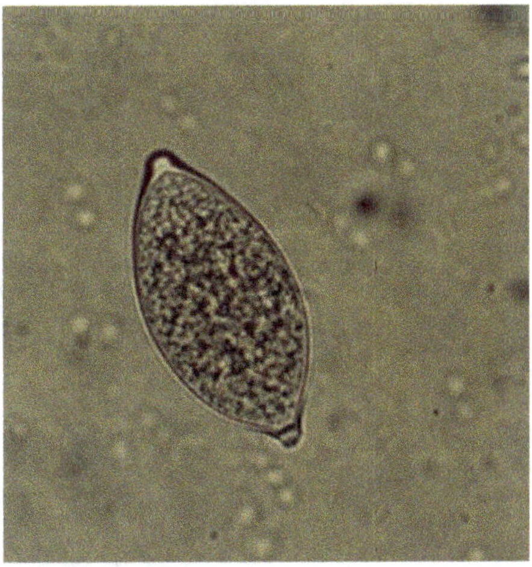

Fig. Symptoms of late blight on leaves

Fig. Symptoms of late blight on stems

Fig. External and internal symptoms of late blight in tubers

The dying, necrotized cells serve as nutrient for the pathogen and under high humidity, a white mildew growth is formed on the underside of the leaves, which represents the sporangiophores and sporangia that emerge through the stomata (Nowicki et al. 2012). During spore formation and germination, large numbers of genes involved in pathogenesis, calcium signaling, and metabolism are upregulated or transcribed in waves (Ah-Fong et al. 2017), whereas genes of the fatty acid biosynthesis pathway are downregulated (Rodenburg et al. 2018). Identification of the proteins and enzymes essential for pathogen growth and development, and linked to the different symptomatic reactions in the plant, can lead to the discovery of potential targets for crop protection chemicals.

High level of moisture is essential for the lesion development, and under optimal conditions the disease can advance very fast and destroy the plant in matter of days (Perez and Forbes 2010).

Impact

The potential economic and social impact of potato late blight is best illustrated by the well-publicized role it played in the Irish Famine in the middle of the nineteenth century. Because of the famine, millions of Irish died or emigrated (Bourke 1993). Other devastating late blight outbreaks have been reported around the world, causing food insecurity, hunger (International Potato Center 2007), and oftentimes crippling the local potato industry. Haverkort et al. (2009) estimated that the global costs and losses due to late blight may take 16% of all global potato production. At 100 €/t the world potato production represents a value of €38 billion today. The 16% loss then represents an annual financial loss of €6.1 billion per annum today, considering that the increase in global potato production in the last decade has mainly been in developing countries, which suffer low yields and the vast majority of these estimated losses to late blight compared to developed countries.

Resistance to Late Blight

After the discovery of the Mexican wild species *Solanum demissum* as an excellent source of resistance, eleven major genes were introduced in cultivated tetraploid potato breeding lines (Black et al. 1953; Malcolmson and Black 1966). Although some of these genes can be considered defeated, others, for example R8, are still effective against current pathogen populations (Vossen et al. 2016). Over 50 R genes have been identified from wild Solanum species as detailed by Rodewald and Trognitz (2013), and the research field remains active with a growing list of genes available for potato breeding programs (Jo et al. 2015; Vossen et al. 2016; Witek et al. 2016; Yang et al. 2017). However, due to crossing barriers and linkage drag, there are only few successful cases where R genes have been introduced into improved tetraploid breeding lines by classical breeding (Bethke et al. 2017). Introduction of a single R gene from wild germplasm is a lengthy procedure as demonstrated by the examples of commercial varieties Bionica and Toluca that contain Rpi-blb2 originating from *S. bulbocastanum*, and were released almost 50 years after the first crosses were made (Haverkort et al. 2016). Genetic engineering bears promise and varieties containing stacked or single R genes are in the process of being released in the markets that accept this technology (Haverkort et al. 2016; Schiek et al. 2016; Pacifico and Paris 2016; Ghislain et al. 2018). As opposed to the major R genes, quantitative resistance was generally expected to be governed by many minor genes. However, recently it was shown that R genes can also have quantitative effects. The potato cultivar Sarpo Mira contains at least four R genes that confer complete resistance against incompatible isolates and a quantitative R gene, Rpi-Smira1 that confers broad-spectrum field resistance (Rietman et al. 2012). A biparental cross using a haploidized resistant clone from the CIP B3 population was used to locate a strong QTL in chromosome 9 (Li et al. 2012a). Subsequent association mapping confirmed the importance of the same genome region for late blight resistance in the tetraploid B3 breeding population (Lindqvist-Kreuze et al. 2014), and recently the R8 gene was identified in the QTL by dRenseq (Jiang et al. 2018).

Identification of new resistance sources and functional resistance or susceptibility genes has been recently greatly accelerated by modern techniques, such as effectoromics and resistance gene enrichment sequencing technologies. To date, all effector proteins identified that are recognized by the plant resistance (R) proteins belong to the RXLR category. Therefore, RXLR effectors cloned into expression vectors have been used to successfully identify functional new R genes from potato germplasm using agroinfection (Vleeshouwers et al. 2011; Vleeshouwers and Oliver 2014). Apoplastic effectors are recognized by pathogen recognition receptors (PRR) and can be used in a similar manner to identify resistance germplasm (Domazakis et al. 2017). In the resistant germplasm, the NB-LRR (nucleotide binding-site leucine-rich repeat) resistance genes can be rapidly identified and cloned using gene-targeted, resistance gene enrichment and sequencing method (Jupe et al. 2013; Witek et al. 2016).

Durability of quantitative resistance will, however, continue to depend on the size of the cultivation area of a variety as well as the dynamics of the pathogen population.

Phytophthora infestans *Populations*

Knowledge on the local pathogen population structure is important for the design of impactful disease management actions (Fry et al. 2015). In recent years, the initiatives EuroBlight (http://euroblight.net/), USABlight (http://www.usablight.org/), and TizonLatino (https://tizonlatino.github.io/) have been carrying out is monitoring of *P. infestans* populations. This work has confirmed that *P. infestans* populations are constantly evolving and novel usually more aggressive genotypes appear periodically replacing the previously dominating genotypes. New genotypes can emerge through divergence from other genotypes, through recombination, or migration from other areas (Knaus et al. 2016). The main mode of reproduction of *P. infestans* is asexual and variable numbers of clonal lineages exist in different countries and regions. Several studies have confirmed that appearance of new genotypes can often be attributed to migration (Fry et al. 2015; Knaus et al. 2016; Saville et al. 2016). Until recently the mating type A1 was dominating worldwide, except in the presumed center of origin, Mexico, where both mating types were found in similar frequencies (Goodwin et al. 1992). This situation has changed dramatically, and A2 has now been reported in Scandinavia and Estonia (Hermansen et al. 2000; Runno-Paurson et al. 2016; Montes et al. 2016), Central Europe (Flier et al. 2007; Li et al. 2012b; Mariette et al. 2016), China (Zhu et al. 2015), Bolivia, Argentina, Uruguay and Brazil (Plata 1998; Deahl et al. 2003; Forbes et al. 1998; Casa-Coila et al. 2017), the USA (Rojas and Kirk 2016), Tunisia (Harbaoui et al. 2014), Algeria (Rekad et al. 2017), India (Chowdappa et al. 2015), and Canada (Danies et al. 2014). However, even though both mating types are present in most cases, no evidence of frequent sexual reproduction has been found, suggesting that the sexual populations are ephemeral (Fry et al. 2015). There are notable exceptions however, such as the Nordic countries, where it was shown that the sexual reproduction in the field is frequent and the oospores surviving in the field over winter in plant debris has led to earlier onset of epidemics (e.g., Widmark et al. 2007). The diversity of *P. infestans* in South and Central America is a particularly interesting question, because these regions are extremely rich in biodiversity of Solanaceous species that are potential alternative hosts of this pathogen and thus can harbor divergent genotypes. Furthermore, the centers of origin of the economically most important hosts, potato and tomato, are there. Interestingly, in South America, no sexual reproduction of *P. infestans* has been reported, and populations maintain strictly clonal structures. In Colombia, Chile, Ecuador, and Peru the A1 mating type has been found mostly (Acuna et al. 2012; Perez et al. 2001; Forbes et al. 1997; Cardenas et al. 2011). In Mexico, in contrast, recombination is frequent and the population is extremely divergent with subdivisions associated with geographic regions (Wang et al. 2017b). Mexico was also shown to be the origin of the current genotypes found in South America and continues to play an important role as the source population of the newly emerged aggressive genotypes in the USA (Saville et al. 2016; Goss et al. 2014). Although *P. infestans* is generally heterothallic requiring two different mating types to form sexual oospores , some isolates are homothallic. Recent

studies have shown that these self-fertile isolates are found more frequently, constituting a new threat to potato and tomato crops because of their increased genotypic variability, better fitness, and greater aggressiveness (Zhu et al. 2016; Casa-Coila et al. 2017).

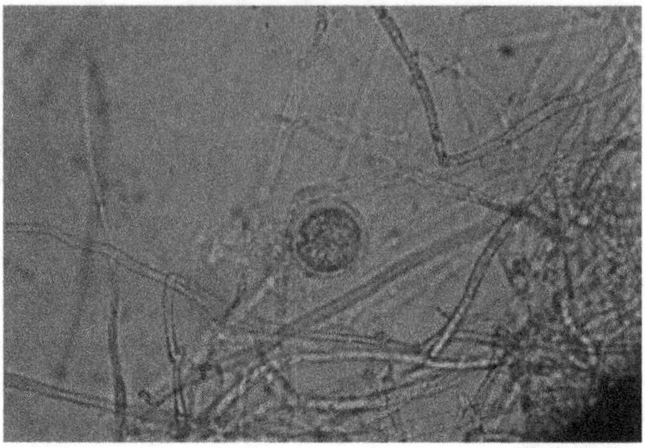

Fig. Oospore of *Phytophthora infestans*

Management

Late blight of potatoes can be suppressed by a combination of approaches. As a polycyclic disease that explodes under favorable conditions, integrated strategies are crucial. These include sanitation measures that eliminate or reduce initial sources of the disease (e.g. infected seed, cull piles, infected neighboring plots, and volunteers), prophylactic fungicide sprays before the appearance of symptoms, curative fungicide sprays and use of resistant cultivars to reduce the rate of disease development, use of early-maturing cultivars to reduce the duration of the epidemic, or planting the crop in seasons or areas where the environment is not favorable for the pathogen.

The application of chemical fungicides continues to be the most common strategy for late blight control, making late blight one of the top drivers for pesticide use in the world. The demand for weekly applications generates a billion-dollar business globally every year (Haverkort et al. 2009). To optimize the use of fungicides, it is important to know the efficacy and type of activity of the active ingredients. The frequency and timing of fungicide applications may depend on the foliar resistance of the cultivar, fungicide characteristics, rate of growth of new foliage, weather conditions, irrigation, and incidence of blight in the region (Cooke et al. 2011). The range of fungicide types available to farmers vary depending on the numbers of products registered and commercialized in their area. In Europe the number of registered products is being reduced due to health and environmental concerns and in some European countries farmers have access to less than ten fungicide products for late blight control (http://www.endure-network.eu). In other countries the process of registration of pesticides is less restricted. For example, in Ecuador, hundreds of fungicide products are registered for late blight control based on more than 30 active ingredients of which most contain old generic substances like mancozeb, cymoxanil, and carbendazim (http://www.agrocalidad.gob.ec), yet in other countries only a couple or so products are readily available to farmers due to trade limitations.

The most efficient and arguably the most elegant strategy to control late blight is the use of host resistance. Today, it is well known that with the use of genetic resistance late blight can be controlled with less fungicide either by lowering the fungicide dose or using longer application intervals (Kirk et al. 2005; Nærstad et al. 2007; Cooke et al. 2011; Liljeroth et al. 2016; Haverkort et al. 2016). The use of resistant varieties could sharply reduce losses from late blight, especially in developing countries, where disease is less well managed for many reasons (e.g. high disease pressure, problems of access to fungicides, and inadequate farmer knowledge of disease dynamics). Nevertheless, the use of resistant varieties continues to be an uncommon disease management approach as susceptible varieties are promoted and required by many wholesalers and processing industries, leaving farmers with little option but to grow susceptible varieties (Forbes 2012). Integrating genetic resistance and chemical control helps in reducing the use of fungicides, decreases production costs, and reduces damage to human health and the environment (Perez and Forbes 2010; Cooke et al. 2011). One way of achieving better design and integration of management elements is through the use of simulation models (reviewed by Forbes et al. 2008) and especially decision support systems (DSS). These typically integrate and organize all available information on the life cycle of *P. infestans*, monitoring of inoculum, weather (historical and forecast), plant growth, fungicide characteristics, cultivar resistance, and thereby predict disease pressure and action thresholds that can guide decision-making. Based on the information provided by the DSS farmers can make informed disease management decisions. DSS can deliver general or very site-specific information to the users via extension officers, telephone, SMS, e-mail, and websites (Cooke et al. 2011). In the case of smallholders in developing countries, basic information to understand the disease is critical to improve management (Nelson et al. 2001; Andrade-Piedra et al. 2009; Ortiz et al. 2019).

Cultural control involves all the activities carried out during agronomic management which alter the microclimate, host condition, and pathogen behavior in such a way that they avoid or reduce pathogen activity (survival, dispersal, and reproduction) (Garrett and Dendy 2001). Among them are the elimination of volunteers and cull piles and associated debris, use of clean seed potatoes preferably certified seed, use of resistant varieties, adequate space between rows and plants, rotation with other crops not susceptible to late blight, adequate hilling, harvest in dry conditions and when the tubes are mature (Garrett and Dendy 2001; Perez and Forbes 2010). Mixtures of potato varieties (resistant and susceptible) offer partial improvement on disease suppression (Phillips et al. 2005; Pilet et al. 2006). Under temperate climate conditions and where tuber infections are a concern, potato vines are typically killed by applying chemical desiccants 2–3 weeks before harvest (Perez and Forbes 2010). Biological control consists of reducing disease through the interaction of one or more live organisms with the disease-causing pathogen or use of extract of plants. Some findings report the use of *Trichoderma* isolates (Yao et al. 2016), *Chaetomium globosum* (Shanthiyaa et al. 2013), *Trichoderma viride* and *Pennicillium viridicatum* (Gupta 2016) and bacteria from the genera *Bacillus, Pseudomonas*, *Rahnella,* and *Serratia* (Daayf et al. 2003) as biocontrol agents in the management of late blight disease in potato. Garlic has, for example, been suggested as a potential intercropping plant for the management of potato late blight disease under Ethiopian condition (Kassa and Sommartya 2006). Still, few biological control measures are used by nonorganic growers due to low efficacy and farmers' lack of knowledge about these options and access to the most efficient products.

Looking Forward

Phytophthora infestans has proven capable of overcoming fungicides and resistant varieties through decades. Potato late blight, thus, continues to be the main potato constraint worldwide despite huge investments in its management. The growing intercontinental trade in potato is also increasing the risk of worldwide dissemination of dominant *P. infestans* strains. The threat from the disease will without doubt continue in the future. Fortunately, encouraging solutions to improve its management are arising from new advances in molecular, sensor, computational, and smartphone technologies. Efficient inoculum monitoring tools are becoming more accessible that can indicate whether *P. infestans* is in an area for the guidance of the initial fungicide spray and *P. infestans* population's movements (Fall et al. 2015). These could even be connected to molecular diagnostics that can predict phenotypic traits of the pathogen population, such as fungicide sensitivity and R-gene interaction, and perhaps even predictions of aggressiveness and fitness in a matter of hours. Spore traps could eventually be mounted on tractors or drones allowing for real-time spatial monitoring (Fall et al. 2015; Fry 2016). New high-throughput methods for monitoring and assessing plant populations are being developed, such as remote sensing, image processing, and web-based farmer-extension service networks. Such indirect disease detection approaches are improving disease surveillance significantly with high potential in remote and low-input production systems. Smartphone-based extension systems linked to rapid and portable molecular diagnostics kits (e.g. Loop-mediated isothermal amplification—LAMP—and lateral flow immuno-diagnostics tools) can support immediate dissemination of disease information to a large number of farmers. If made widely available such smartphone tools, hand-held portable diagnostics kits and novel weather sensors could lead to precision management and significant impacts on production capacity. Farmer-extension service networks based on internet connectivity are already important options that improve disease surveillance and supplying fast access to appropriate and up-to-date knowledge on pathogen distribution and management (Fry 2016). Capacity building (for farmers and extension services) to improve disease management skills, coupled to early-maturing and resistant cultivars, novel diagnostic tools, improved DSSs, and low-toxicity fungicides have the potential to reduce crop loss, management costs, and environmental impact, and even more so as biological options become accessible to farmers.

Current research opportunities in late blight management focus on *P. infestans* studies, fungicide testing, dissemination of resistant cultivars and validation of DSSs and training. In addition to continued research and extension efforts, alliances with the agro-chemical industry seem to be necessary to fully achieve integrated pest management strategies (Pacilly et al. 2016). This is being promoted through local and regional research networks that via collaboration strengthen institutional capacity in research and extension related to late blight (see http://euroblight.net/). Emerging research topics on *P. infestans* include the role of the pathogen–microbiota interaction in promotion or suppression of the disease

(Larousse and Galiana 2017) as well as the metabolism of *P. infestans* (Judelson 2017), its effector repertoire on the plant, and, *in fine*, how it promotes or suppresses the disease. At the same time, *P. infestans* as an infectious entity is no longer only considered at the species or lineage level but also at the level of a resident microbiota or part thereof (Larousse and Galiana 2017).

Early Blight

Hans Hausladen and Birgit Adolf

Symptoms

Foliar symptoms of early blight (EB) are dark brown to black necrosis. First foliar symptoms become visible on the lowest and therefore oldest leaves just a few weeks after emergence. Initially, dark brown dot-like blotches appear, a few millimeters in diameter. The necrotic area gradually increases and the leaf symptoms grow to take up the whole of the green leaf tissue. Often, the lesions are restricted by leaf veins and take on an angular shape. The size of the necrosis can vary in width, from a few millimeters to 2 cm. Within larger lesions, a series of dark concentric rings are visible. This target pattern is typical of EB symptoms. Subsequently, the necrotic leaf tissue is often surrounded by a chlorotic border caused by fungal mycotoxins, which turn the leaf tissue yellow. The chlorosis can extend to the whole infected leaf. During EB progression the infected areas enlarge, and the whole leaf becomes necrotic and falls off. In Europe, a heavy increase of EB infestation occurs from mid-July onwards, especially during hot and dry weather or when the potato crop is under stress, and on physiologically older plants. EB then starts spreading from the lower leaves to the middle and finally upper leaf levels.

Alternaria conidia which are washed of the leaves can also infect tubers. The symptoms of EB on tubers are dark, slightly sunken lesions. The dry or

Fig. Initial (left) and advanced (right) symptoms of potato early blight in foliage

Fig. Symptoms of potato early blight in tubers

hard rot of tubers causes storage losses, reduces quality of table potatoes, and reduces germination capacity of seed potatoes.

It isn't possible to distinguish between the different *Alternaria* spp. causing EB based on the symptoms even though sometimes symptoms referred to as brown leaf spot (small, irregular to circular, dark brown spots ranging in size from a pinpoint to 4 mm) are attributed to *A. alternata* (Fairchild et al. 2013).

Causal Organism

The main causal agent of early blight on solanaceous crops is generally considered to be *Alternaria solani* Sorauer (Gannibal et al. 2014). However, there are reports of other large-spored *Alternaria* spp. involved in EB of potato. Rodrigues et al. (2010) found that *A. grandis* Simmons, but not *A. solani*, was the causal agent of EB affecting potato plants in several growing regions in Brazil, and Duarte et al. (2014) confirmed that this species can cause EB on potato in field trials using artificial inoculation with *A. grandis*. *Alternaria protenta* has been detected as the causal *Alternaria* spp. for EB in Algeria (Ayad et al. 2017) and was, together with *A. grandis* and *A. solani*, found to be part of the complex of *Alternaria* spp. detected on EB lesions in Belgium (Landschoot et al. 2017a). Of the small-spored *Alternaria* spp., *A. alternata* occurs on EB lesions on a regular basis, but is considered as a secondary invader (Leiminger et al. 2014; Stammler et al. 2014; Rotem 1994).

Alternaria solani overwinters as mycelium, chlamydiospores or conidia in the soil and infested plant debris (see disease cycle in Fig.). In spring, the primary infection occurs through inoculum (conidia) carried to the lower leaves by rain splashes. The pathogen is able to penetrate the leaf tissue directly through the intact epidermis or through stomata and wounds. First symptoms of EB on leaves become visible 2–4 weeks after the emergence of the potato crop. Initially, the older leaves closer to the ground are infested. The fungus is restricted to the lower leaf level for several weeks and seems initially to be of no concern. Conidia formation occurs on

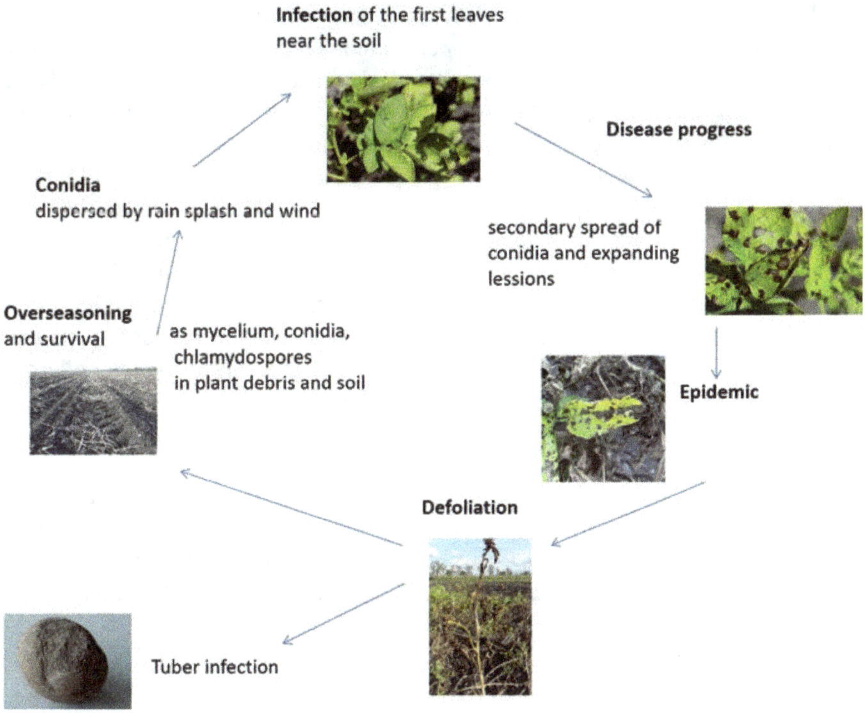

Fig. Disease cycle of potato early blight

the necrotic leaf tissue at temperatures between 5 and 30 °C (optimum 20 °C). This secondary inoculum is disseminated by wind and causes infections of the surrounding leaves and stems. The latency period averages 3–7 days. During favorable infection conditions, and once the plant has got to a certain age, *A. solani* very rapidly colonizes the middle and upper leaf layers. In the field, a cascade-like progression of the fungus from the lower, via the middle, to the upper leaves is visible. Heavily infested leaves fall off and remain as inoculum source on and in the soil.

The disease progression of EB depends on weather parameters, plant age/crop growth stage, maturity group, susceptibility of the cultivar and inoculum concentration, the latter being influenced by short crop rotations.

Favorable weather conditions for infection with *A. solani* are temperatures above 22 °C and leaf wetness for more than 8 h. Furthermore, alternating wet and dry periods favor disease development.

An important factor for disease progression is the crop growth stage. In addition to the cascade-like spread within plants upwards from older to younger leaves, older plants are significantly more susceptible due to the earlier planting emergence time.

Tuber infestation is caused by conidia that are washed into the soil. The fungus can enter the tuber tissue through lenticels or mechanical injuries of the skin during harvest.

EB can also occur on other host plants apart from potatoes. It has been observed on hosts such as tomato (*Solanum lycopersicum* L.), eggplants (*S. melongena* L.), hairy nightshade (*S. sarrachoides* Sendt), black nightshade (*S. nigrum* L), horse nettle (*S. carolinense* L.), pepper (*Capsicum* spp.), and non-solanaceous weeds (Jones et al. 1993; Pscheidt 1985; Hausladen and Aselmeyer 2017).

Differentiation of *Alternaria* species within the large-spored or the small-spored group based on morphological traits is time-consuming and requires experience. The molecular approach to delineate large-spored isolates fast and precise is a multi-locus analysis based on different partial gene regions, like the internal transcribed spacer regions 1 and 2 and intervening 5.8S nrDNA (ITS), glyceraldehyde-3-phosphate dehydrogenase (GAPDH), RNA polymerase second largest subunit (RPB2), translation elongation factor 1-alpha (TEF1), the Alternaria major allergen gene (Alt a 1) (Woudenberg et al. 2014). Based on the RPB2 and calmodulin gene sequence *A. solani* can be distinguished from A. grandis and A. protenta (Landschoot et al. 2017a). For the small-spored species the glyceraldehyde-3-phosphate dehydrogenase (gapdh), RNA polymerase second largest subunit (rpb2), translation elongation factor 1-alpha (tef1), Alternaria major allergen gene (Alt a 1), endopolygalacturonase (endoPG), an anonymous gene region (OPA10-2), and the histone H3 gene can be used for differentiation (Woudenberg et al. 2015; Landschoot et al. 2017a).

Impact

After potato late blight (*Phytophthora infestans*), EB represents one of the most important fungal diseases of potato today. It can be found in most potato-growing countries (Woudenberg et al. 2014), but *A. solani* is described as a very important pathogen especially in warmer areas due to the requirement for higher temperatures. In susceptible potato cultivars in particular, as well as in locations with increased occurrence of *A. solani*, the disease can cause considerable yield losses. The potential EB-induced damage for different countries is shown in Table.

The necrosis of leaf tissue and considerable defoliation caused by *Alternaria* infestation reduces the assimilation area of the potato plant, and therefore has a

Table Yield loss due to early blight in different regions

Country	Yield loss
Australia	>20% (Horsfield et al. 2010)
Brazil	up to 58% (Campo Arana et al. 2007)
Germany	2– >40% (Leiminger and Hausladen 2014)
Poland	6–34% (Kapsa 2004)
Israel	up to 24% (Shtienberg et al. 1996)
United States	18–39% (Harrison and Venette 1970)
South Africa	20–50% (Van der Waals et al. 2001)

negative effect on tuber size and starch content. *Alternaria solani* is also able to infect tubers. The subsequent hard rot of the tubers leads to reduced quality for marketing as table and processing potatoes. In Europe, the occurrence of tuber infections is known only in exceptional years.

Pathogen Populations

The genetic diversity among *A. solani* isolates is relatively high for an asexually reproducing fungus. Furthermore, when analyzed by virulence assays (VC), random amplified microsatellites (RAMS), random amplified polymorphic DNA (RAPD), and AFLP marker techniques, there is no clear clustering of isolates according to geographical origin, year, or even deriving from the same field (Van der Waals et al. 2004; Leiminger et al. 2013; Odilbekov et al. 2016). Polymorphism exists even within the usually strongly conserved mitochondrial DNA. Concerning the cytb gene, Leiminger et al. (2014) detected two genotypes in German *A. solani* populations which differed in their intron–exon structure. The occurrence of the two genotypes was confirmed for *A. solani* populations in the USA (Bauske et al. 2018), Belgium (Landschoot et al. 2017b), and Sweden (Odilbekov et al. 2016).

The registration of *Alternaria*-specific fungicides with single site modes of action for potato changed the population structure of *A. solani* in many countries due to the occurrence of point mutations, leading to reduced sensitivity of mutated isolates. For Quinone outside inhibitor fungicides (QoIs) like Azoxystrobin, the change of amino acid phenylalanine to leucine at position 129 in the cytb gene (F129L) has been demonstrated to be the reason for reduced sensitivity compared to wild-type isolates (Pasche et al. 2005). The presence of the F129L mutation in *A. solani* populations has been shown in different countries such as the USA (Pasche et al. 2004), Germany (Leiminger et al. 2014), Sweden (Odilbekov et al. 2016), and Belgium (Landschoot et al. 2017b). The loss in sensitivity towards succinate dehydrogenase inhibitor fungicides (SDHIs) like Boscalid can be caused by different point mutations, as the SDH is composed of four subunits and the binding site of the fungicides is formed by three of them (subunits B, C, D). For *A. solani* populations, five possible mutations have been described so far: H278Y and H278R in subunit B, H134R and H134Q in subunit C, and D123R and H133R in subunit D (Mallik et al. 2014; Metz et al. 2019). Landschoot et al. (2017b) describe the presence of two different SDH mutations in one isolate. Frequently, isolates carry both the F129L mutation in the cytb gene and one of the SDH mutations (Landschoot et al. 2017b).

The sensitivity of baseline isolates to demethylation inhibiting fungicides (DMIs) can vary substantially, but nonbaseline isolates remain sensitive, whereas a distinct loss in sensitivity to anilino-pyrimidine (AP) fungicide pyrimethanil exists. The primary mode of action has not yet been discovered for AP chemistries; therefore the resistance mechanism is currently unknown for this established fungicide group (Fonseka and Gudmestad 2016).

Resistance against QoIs in *A. alternata* populations is caused by the G143A mutation in the cytb gene (Ma et al. 2003). Mutations conferring reduced sensitivity to SDHIs for this Alternaria species are the H277R/Y in SDH subunit B, H134R in subunit C, and D123R and H133R in subunit D (Avenot et al. 2008, 2009). Fairchild et al. (2013) detected isolates resistant to pyrimethanil and Fonseka and Gudmestad (2016) stated high DMI EC50 values in baseline isolates.

Management

Integrated pest management to control early blight requires the implementation of several approaches. The disease is primarily controlled by the use of cultural practices (to reduce the soil born inoculum), less susceptible cultivars and the use of pesticides.

Phytosanitary aspects One of the main components in this case is the crop rotation, which influences the occurrence of early blight. The fungus *A. solani* persists as mycelium or spores in plant debris or soil in the field from one potato-growing season to the next. Therefore crop rotation, including the control of host plants such as weeds (black shadow) in the nonhost crops, reduces the initial soil born inoculum. A short crop rotation with host crops (tomato, potato) results in an earlier and more severe early blight epidemic (Shtienberg and Fry 1990). In addition, the removal or burning of infected plant debris reduces the inoculum level.

A further option to reduce the primary inoculum in the soil is biofumigation. Biofumigation means the suppression of soilborne pathogens by isothiocyanates (ITCs), which derive from hydrolization of glucosinolates by myrosinase in disrupted plant cells. Biofumigant plants (e.g. white mustard, leaf radish) can reduce the early blight disease progression in the crop (Volz et al. 2013).

Tuber harvest and storage *Alternaria solani* also can infect the potato tuber. The fungus cannot infect through the intact periderm, and so the risk of tuber infection can be reduced by allowing tubers to fully mature before harvest. Avoiding wounding at harvest and providing storage conditions to promote wound healing can also reduce tuber infection (Venette and Harrison 1973).

Pathogen-free seed The use of disease- and virus-free seed potatoes is the basis for an economical potato production. Virus-infected potato plants are more susceptible to early blight than healthy plants.

Biotic and abiotic stress Potato plants stressed by biotic or abiotic factors are more susceptible to early blight disease compared to nonstressed plants. There are different types of abiotic stress for plants during the growing season, such as drought, high temperature, and overhead irrigation. Additionally, overhead irrigation can prolong the leaf wetness period, allowing successful fungal infection.

Biotic stress is driven by insects (e.g. aphids, Colorado beetle), which are also known to be virus vectors.

Plant nutrition A balanced nutrition for the potato plants during the growing period is the basis for optimal plant growth and potato yield. Ideal soil fertility and plant nutrition can decrease the severity of *A. solani* (Lambert et al. 2005; MacDonald et al. 2007). Under specific conditions, such as drought when plants are unable to get enough nutrients from the soil through the roots, a foliar fertilizer can reduce the nutrient deficiency and reduce plant susceptibility to early blight.

The fertilizer form can also influence the disease progression of *A. solani*. The use of calcium cyanamide results in a delay of early blight disease, as the fungicidal side effects of degradation products of calcium cyanamide can reduce the initial inoculum in the soil (Volz et al. 2013).

Resistant cultivars Cultivars with reduced susceptibility to early blight are available; however, no completely resistant genotypes have been found so far. The observed field resistance of varieties to foliage infection is associated with plant maturity. Early maturing cultivars are in general more susceptible, and late maturing cultivars are more resistant to *A. solani* (Johanson and Thurston 1990; Abuley et al. 2017). There is no correlation between maturity group and the occurrence of the first early blight symptoms on the leaves, but a strong correlation between maturity group and disease progression. The epidemic in early maturing cultivars starts earlier. Interestingly, there are varieties within a maturity group which are more resistant to early blight (Johanson and Thurston 1990; Leiminger and Hausladen 2014). Overall, there is a possibility of influencing disease progression by planting more resistant cultivars.

Use of fungicides The most common method for controlling early blight in potatoes is the use of chemical pesticides. Some fungicides which are used for the control of late blight (Phytophthora infestans) also have some effect on Alternaria solani (e.g. maneb, mancozeb, chlorothalonil, triphenyl tin hydroxide). The most effective fungicides for control of early blight contain active ingredients from the strobilurins group, and azols.

Strobilurins, also known as Quinone outside inhibitors (QoIs), are an important class of fungicides in agriculture because they have a broad-spectrum activity. They inhibit mitochondrial respiration in fungi by binding to the Qo site of the cytochrome b (cytb) complex, blocking electron transfer and inhibiting ATP synthesis (Bartlett et al. 2002).

Carboxamides (SDHI) inhibit the enzyme succinate dehydrogenase (Sdh), a component of complex II in the mitochondrial electron transport chain (Kuhn 1984). Despite the two groups of fungicides having a similar mode of action, SDHI and QoI fungicides show no cross-resistance.

The triazol group belongs to the DMI fungicides, which inhibit one specific enzyme, C14-demethylase. This enzyme plays an important role in ergosterol biosynthesis. Ergosterol is important for functional membrane structure and for the development of functional cell walls.

Due to the single site mode of action of strobilurins (QoI) and carboxamides (SDHI), these fungicides have a high risk of development of resistance. In several potato-growing areas, mutants are found which show a reduced sensitivity in in vitro and in vivo trials and also a reduced efficacy in the field (see Pathogen populations).

Most of these fungicides have a very limited curative activity and should be used preventively. EB control is mainly achieved by multiple and frequent application of protectant fungicides. The optimization of fungicide usage for the control of early blight is a considerable challenge. The fungus produces huge amounts of secondary inoculum during the growing season. Therefore different DSS (decision support systems) are available to optimize the use of fungicide applications for the management of early blight in potato. One possibility is to use threshold values based on the disease progress (Leiminger and Hausladen 2012). In some countries, disease management is based on interactive computer-based systems dealing with forecasting favorable weather conditions for infection by Alternaria solani or temperature degree-day thresholds. Alternatively, the recommendation for the use of fungicides is based on plant development (plant size, plant age, onset of potato flowering).

Overall, the combination of plant age and host resistance, disease progress, and calculated weather-based infection risk is used as the basis for an integrated pest management.

Looking Forward

The control of early blight will be a considerable challenge in the future. The increasing importance of early blight in potatoes is due to a number of factors. Climatic change and global warming will result in more conducive conditions for the infection, growth, and disease progress of the fungus in several potato-growing areas. Increasing temperatures during the growing season result in favorable conditions for the pathogen, due to its requirement for higher temperatures. More abiotic stress (drought) increases the susceptibility of the plant, and will therefore also promote disease progress.

In addition, integrated pest management based on continuous pesticide treatments will become more and more ineffective. The occurrence of mutants towards different modes of action will result in reduced efficiency in the control of EB by fungicides. Therefore in the future, successful control of EB requires the implementation of cultural practices, the use of pesticides with a focus on fungicide resistance development, and the cultivation of less susceptible cultivars.

Wart

Jaroslaw Przetakiewicz

Causal Organism

The pathogen causing potato wart disease, *Synchytrium endobioticum* (Schilb.) Perc., was first discovered by Schilberszky in Hungary. In fact, the pathogen was known firstly in Europe. In 1876, potato wart disease was found for the first time in the UK (Hampson 1993; Flath et al. 2014). In the older classification this species has been included to Protista Kingdom. Nowadays, *Synchytrium endobioticum* (Schilb.) Perc., belongs to Fungi Kingdom, phylum Chytridiomycota, order Chytridiales, family Synchytriaceae, genus *Synchytrium*, and species *endobioticum*. The genus *Synchytrium* included about 200 species which are endobiotic halocarpic organisms that have inoperculate sporangia. All species of the genus *Synchytrium* are parasites but the most important economically and phytosanitary is *S. endobioticum* the causal agent of potato wart disease. The disease is also known by various common names like black wart, cauliflower disease, warty disease, potato tumor, potato cancer, black cancer, or black scab. *S. endobioticum* is an obligate soil-borne biotrophic fungus which is considered to be the most important worldwide quarantine plant pathogen of cultivated potato. Cultivated potato (*Solanum tuberosum*) is the primary host, but the fungus, under experimental conditions, can also infect wild species in genera *Capsicastrum*, *Duboisia*, *Hyoscyamus*, *Lycium*, *Nicotiana*, *Nicandria*, and *Physalis* (Obidiegwu et al. 2014). The pathogen is a primitive fungus which stimulates its host to produce hypertrophic outgrowths on young potato organs, such as eyes, sprouts, young tubers, stolons, stems, leaves, and even flowers but never roots. The fungus does not form hyphae but forms sporangia that produce about 200–300 motile zoospores (Obidiegwu et al. 2014). After infection, *S. endobioticum* produces two different kinds of sporangia in the galls. Summer sporangia have a thin cell wall and form haploid zoospores which are emerging and steady reinfection of the host tissue like sprouts, tubers, eye tubers, stolons, and roots (only in tomato) (Przetakiewicz 2014a). In appropriate conditions after isogamy of haploid zoospores to diploid zygotes which are able to infect host cells and form winter sporangia which are embedded deeper into the host tissue than the sori (always on the surface). Winter (resting) sporangia are the dormant structures by which the fungus disperses to establish new infections. They are usually spherical to ovoid in shape and 24–75 μm in diameter with thick-walled (triple wall) structure, which is ornamented with irregularly shaped wing-like protrusions. The spores can survive for a long time without plant hosts. After 43 years, in favorable conditions, disease may develop even from single spores of *S. endobioticum* (Przetakiewicz 2015b, 2016).

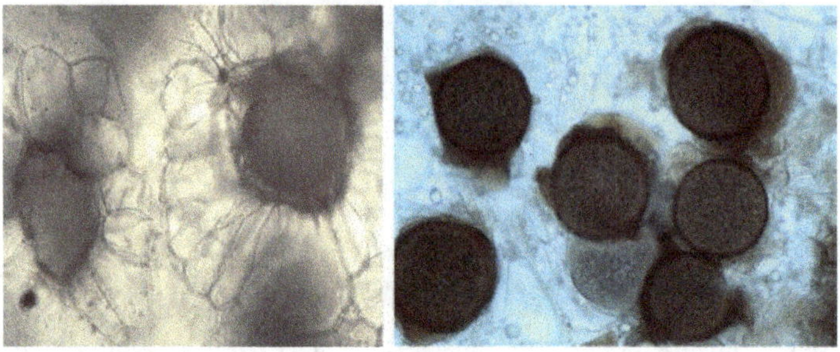

Fig. Summer (left) and winter (resting) sporangia (right) of *Synchytrium endobioticum*. Thin-walled summer sporangia of pathotype 3(M1) are enclosed within one membrane forming sorus on surface of wart tissue of cultivar Asche Sämling. Thick-walled winter sporangia of pathotype 8(F1) in wart tissue of cultivar Sonda

Fig. Symptoms of potato wart disease above ground (left) and underground (right). The galls on stem and foliage above ground are colored green, while subterranean warts on stem base, stolons, and tubers are colored white. Pathotype 2(Ch1) on cultivar Cykada (left). Pathotype 18(T1) on cultivar Deodara (right)

Symptoms

Usually, the symptoms of potato wart disease are not visible on plants, although there may be a reduction in plant vigor. Only in very suitable conditions small greenish warts might be visible on the top of plants: stem, foliage and in extremely conditions on inflorescences. (Obidiegwu et al. 2014). In the most cases the symptoms are visible on underground parts of potato on stolons, stems, bulbs, eye of matured tubers. In soil with the high content of winter sporangia (above 500/g) may lead to infection all eyes of seed potato and develop only warts without any emergences of potato (Przetakiewicz 2014a). The typical symptoms of the disease on tubers are the proliferating warts which may vary markedly in form but are primarily spherical to irregular. The warts are usually whitish or green if exposed to light, but gradually darken eventually rot and disintegrate. Rarely, the warts can be yellow or purple to brown. The color is dependent on potato variety. If variety produces purple sprouts, then the warts will be purple too. The warts are similar to cauliflower and they usually have the same color but sometimes they are compared to walnut kernel (Przetakiewicz 2014a). The warts are proliferated to gall. Galls vary markedly in form but are primarily spheroid. They are primarily parenchymatous and phytoteratological and not phytooncological. Although, the disease is described as potato tumor or cancer as well as black cancer they should not be referred to as tumors (Hampson 1993). Warts differ in size from pea-sized nodules to the size of a fist. The warts maintain in lab condition (for inoculum production) can reach 220 g from one eye/sprout. At maturity the galls become colored black and lead to total tuber decay (Przetakiewicz 2014a).

Simultaneous germination of all buds in one eye results in wart-like outgrowths very similar to those caused by *S. endobioticum*. However, these pseudo-warts consist of abundant pointed shoots compacted together. No winter sporangia are present in the affected tissue. Symptoms of powdery scab caused by *Spongospora subterranea* f.sp. *subterranea* or potato smut caused by *Thecaphora solani* can be mistaken for wart occurrence. A view under the microscope reveals that spore balls look different from winter sporangia of *S. endobioticum*. The size of warts depends not only on environmental conditions (cool summer, wet soil, etc.) but on potato variety. On extremely susceptible variety specified pathotype of *S. endobioticum* can differ markedly to very big size producing winter sporangia in the last stage of growing. On the slightly susceptible variety specified pathotype of *S. endobioticum* is able to produce very small warts or only to influence on weakly proliferation of host tissue but producing winter sporangia in the first stage of infection. The spores were visible after 13 days after inoculation of slightly susceptible varieties. In uncomfortable conditions (high temperature or dry soil) warts stop growing and begin to produce winter spores (Przetakiewicz 2014a).

Impact

Potato wart disease is so important that, for some 65 years, quarantine and domestic legislations have been in force throughout the world to prevent its spread (Anon 2015). The economic impact of disease caused by this pathogen is not only from disease losses but from loss of international trade markets, long-term quarantines, and regulatory restrictions placed on infested areas and the buffer zones (Przetakiewicz 2014a). Chemical control of *S. endobioticum* is not possible. The only strategies to confine the disease are strict quarantine and phytosanitary measures as well as cultivation of resistant cultivars (Obidiegwu et al. 2014). The availability of resistance cultivars allowed governments to issue regulations prohibiting the cultivation of susceptible cultivars (Baayen et al. 2006). For example, since 1955 only cultivars resistant to pathotype 1(D1) could be registered and grown on the Polish territory. As stated in the Food and Veterinary Office Mission Report, no potato plants or tubers with symptoms of *S. endobioticum* pathotype 1(D1) have been detected in Poland since the 1950s–early 1960s. This is a result of growing only resistant cultivars to mentioned pathotype of *S. endobioticum* (Przetakiewicz 2008). Worldwide prevention is based on the control of disease spread and pathogen exclusion via regulatory action. *S. endobioticum* has a very limited capacity for natural spread, which is principally why it has been possible to control it so effectively by statutory means. Nevertheless, *S. endobioticum* is a classic example of the distribution of plant pathogens by man. Regulatory action has largely restricted the spread of the disease within potato-growing regions, as the seriousness of the disease was quickly recognized (Hampson 1993). Once *S. endobioticum* has been introduced into a field, the whole crop may be rendered unmarketable and moreover the fungus is so persistent that potatoes cannot be grown again safely for many years, nor can the land be used for any plants intendent for export. Based on the results obtained by Przetakiewicz (2015b), winter sporangia of *S. endobioticum* are very persistent and capable of retaining viability for as long as 46 years. The single spores from the inter-host period were still infective after 43 years. These results should be taken into consideration when de-scheduling previously infested plots even after 40 years or longer, especially in the mountainous areas. *S. endobioticum* is still of great economic importance in cool areas and wet mountainous regions. The detection of potato wart disease on Prince Edward Island during the 2000 growing season resulted in an estimated $30 million loss to the island's economy in the first year.

Pathogen Populations and Distribution

Potato wart disease appears to have arisen in the potato-growing area of Andean South America (Hampson 1993). Wart-like outgrowths on early Peruvian tuber-shaped pottery were interpreted as potato wart disease. The disease is likely to have arrived in Europe from South America because the Great Potato famine of the 1840s in Ireland induced European growers to import potato germplasm from South America. The introduction of the pathogen to Europe was possible by diseased tubers, infested soil or contaminated bags along with shipments of guano (Hampson 1993). *S. endobioticum* spread at the end of the nineteenth century from the center of origin in the Andes first to Europe and North America and subsequently across whole potato-growing areas of Asia, Africa and Oceania. Historic account has it that potato wart disease entered England in 1876 or 1878 while another view upholds that the disease has been present in the Liverpool province of England in 1876 or 1878 (Obidiegwu et al. 2014). In 1901, the disease was officially recognized in the UK. Potato wart disease and its causal agent were described by Schilberszky who received in 1888 warted tubers of cultivar Maercker-Zwiebel that had been grown locally from seed tubers imported from England. It spread widely in Europe, but statutory measures finally restricted its distribution and it has spread only to a limited extent to other parts of the world. According to EPPO Pest Quarantine Database

(Anon 2015), *S. endobioticum* occurs locally in almost all EPPO countries. The distributions are fragmentary as a result of statutory control. According to national reports, it has been found but is not established in Belgium, France, Luxembourg as well as in Lebanon. Found in the past but eradicated in Portugal (unconfirmed). In Asia countries *S. endobioticum* occurs in Armenia, Bhutan, China, India, Georgia, Nepal, Turkey. In Africa: South Africa, Tunisia, Algeria, Zimbabwe, and Egypt (absent, unreliable record). North America: Canada, Mexico (absent formerly present), USA (eradicated). South America: Bolivia, Chile (eradicated), Ecuador, Falkland Islands, Peru, Uruguay (absent, confirmed by survey). Oceania: New Zealand (South Island). Numerous pathotypes of the fungus occur and are defined by their virulence on differential potato cultivars. In Europe, more than 40 pathotypes of *S. endobioticum* have already been identified (Obidiegwu et al. 2014; Przetakiewicz 2014b, 2015a). A pathotype is characterized by its pattern of virulence or avirulence to a series of differential cultivars of potato. Pathotypes 1(D_1), 2(G_1), 6(O_1), 8(F_1), and 18(T_1) are the most relevant in Europe. Other pathotypes occur mainly in the rainy mountainous areas of central and eastern Europe (Alps, Carpathians). They persist mainly in small garden potato plots and not in commercial potato crops (Przetakiewicz 2014a). Mitochondrial genomic variation shows that *S. endobioticum* has been introduced into Europe multiple times, that several pathotypes emerged multiple times, and that isolates represent communities of different genotypes (van de Vossenberg et al. 2018a).

Resistance

The biggest discovery was finding resistant cultivars of potato to *S. endobioticum* among cultivated ones. Systematic studies on resistance of potato cultivars to *S. endobioticum* started in England in 1909. Resistance sources to *S. endobioticum* were found in old cultivars such as Snowdrop and Flourball, which facilitated resistance breeding. Conventional breeding programs were successful in controlling potato wart disease through the development of resistant cultivars early in the twentieth century (Obidiegwu et al. 2014). Breeding for resistance was successful, thanks to the arability of a dominant gene that blocked development and reproduction of originally introduced pathotype 1(D1) of *S. endobioticum*. Unfortunately, in Europe wart development on resistant potato cultivars was first discovered in 1941 in Germany [pathotype 2(G1)] and former Czechoslovakia [pathotype 3(S1)]. The new pathotypes have been proved to be difficult to control and eradicate then the original pathotype 1(D1) (Baayen et al. 2006). Molecular mapping studies provide evidence that wart resistance to pathotype 1(D1) of *S. endobioticum* can be conferred by a single locus from different sources. Hehl et al. (1999) mapped the single dominant gene *Sen 1* for resistance to pathotype 1(D1) in diploid mapping population. The gene *Sen 1* is located on potato chromosome IX. Brugmans et al. (2006) also used a diploid potato linkage map to locate *Sen1-4*, a second dominant gene for resistance to pathotype 1(D1). This gene is located on the long arm of chromosome IV. In these two mentioned populations, the resistance segregated as a monogenic trait. Ballvora et al. (2011) discovered the first loci for virulent pathotypes 2(G1), 6(O1), and 18(T1) in two tetraploid half-sib families, in which the resistance to pathotype 1(D1) also segregated. In contrary to earlier studies in diploid populations (Hehl et al. 1999; Brugmans et al. 2006), the phenotypic distribution of wart resistance appeared quantitative in the two mapping populations analyzed (Ballvora et al. 2011). The quantitative resistance locus (QRL) *Sen2/6/18* on chromosome I expressed resistance to pathotypes 2(G1), 6(O1), and 18(T1). The QRL *Sen18* on chromosome IX expressed resistance to pathotype 18(T1). And the third QRL Sen1 on chromosome XI expressed resistance mainly to pathotype 1(D1) (Ballvora et al. 2011). Groth et al. (2013) mapped quantitative trait loci (QTL) for resistance to pathotype 1(D1), 2(G1), 6(O1), and 18(T1). The QRL for all four pathotypes were located on chromosomes II, VI, VIII, and IX, and QRL for pathotypes 2(G1), 6(O1), and 18(T1) on chromosomes VII and X. The QRL detected in this study were different from the ones in Ballvora et al. (2011). The cultivar Panda used in this study (Groth et al. 2013) as a resistant parent to pathotype 1(D1), 2(G1), 6(O1) and 18(T1), is in fact slightly susceptible to all four pathotypes. These results were confirmed in CORNET project (acronym SynTest) (Przetakiewicz, unpublished results 2014) Consequently, breeding is hampered by a lack of dominant major genes for resistance to virulent pathotypes. The wart resistance has several sources and diverse backgrounds which resulted in a few potato cultivars resistant to virulent pathotypes of *S. endobioticum*. Nevertheless, recent results indicated for identify a newly locus *Sen2* located on chromosome XI which provides resistance to at least seven various virulent pathotypes of *S. endobioticum* (Plich et al. 2018).

Management

The first acts were issued in 1908 in Ireland, and a few years later in Scotland, England, and Germany. The first legislation related primarily to prohibit the import from other countries of potatoes infected by *S. endobioticum.* In that time the acts did not take into account the possibility of the spread of the disease within the country. It was a reason of increasing number of outbreaks in such a short time in many countries. In consequence of the wide spread of potato wart disease was the development of research on the biology controlling of the fungus. EPPO includes *S. endobioticum* into the A2 quarantine list because of its long persistence in soil and the severe losses it inflicts to potato crops. The United State Department of Agricultural (USDA) has the fungus on its official list of selected agents and toxin. The European Union issued a specific requirement in the Council Directive 69/29/EC of 8 December 1969 on control of Potato wart disease and the Council Directive 2000/29/EC of 8 May 2000 on protective measures against the introduction into the Community of organisms harmful to plant or plant products and against their spread within the Community (Obidiegwu et al. 2014).

Looking Forward

Although *S. endobioticum* seems to be in remission in Europe, new foci have appeared in countries with warmer continental climate (Turkey, Georgia, Bulgaria, and Greece) (Anon 2015; Gorgiladze et al. 2014; Vloutoglou pers comm. 2015). This may suggest adaptation of *S. endobioticum* to warmer and/or dryer climate as well as the lack of adequate controls in countries where the disease has not been present before. Moreover, new pathotypes have been discovered recently (Çakir et al. 2009; Przetakiewicz 2015a). It can be expected that new pathotypes of *S. endobioticum* might appear, since there are many reasons for new pathotypes to arise (Melnik 1998). The newest Pathotype 39(P_1) was detected in the rainy mountainous area, in small garden potato plots, when the old traditional cultivars of potato are cultivated without any crop rotations. It seems therefore that where climatic conditions are suitable for *S. endobioticum* to take place and the growing of slightly susceptible cultivars is possible, the development of new pathotype is favored (Przetakiewicz 2015a). Although, local pathotypes persist mainly in small garden potato plots in economically unimportant potato-growing regions, they are still very important for quarantine and phytosanitary measures as spores can spread in rain water (Przetakiewicz 2015b). Phenotypic assessment of resistance to *S. endobioticum* is laborious and, time-consuming. Diagnostic DNA-based markers closely linked with or, even better, located within wart resistance genes would greatly facilitate the early detection and combination of different resistance sources and are therefore highly desirable (Obidiegwu et al. 2014). Natural DNA variation in wild and cultivated potato germplasm provides an excellent platform for the discovery of diagnostic tools for marker-assisted selection and resistance gene cloning (Obidiegwu et al. 2014). A new dominant gene *Sen2* on chromosome XI provides extreme resistance to pathotypes 1(D1), 2(G1), 2(Ch1), 3(M1), 6(O1), 8(F1), 18(T1), and 39(P1). In the future, this gene will offer potentials for the efficient selection of new commercial cultivars that are resistant to multiple *S. endobioticum* pathotypes. EPPO Standard PM 7/28 (2) (Anon 2017) recommends various biotests using differential potato cultivars for the identification only 4 pathotypes [1(D1), 2(G1), 6(O1) and 18(T1)]. Pathotype determination is labor-intensive and time-consuming too, especially for the identification of local pathotypes which require more differential cultivars (Przetakiewicz 2017; Przetakiewicz and Plich 2017). Molecular diagnostic tools (TaqMan PCR method) are currently available for the identification of pathotype 1(D1) and its discrimination from non-1(D1) pathotypes (Bonants et al. 2015; van de Vossenberg et al. 2018b). However, this method requires costly probes and cannot be used for the identification of virulent pathotypes. The recent report revealed no sequence polymorphisms between the five *S. endobioticum* pathotypes, indicating that *S. endobioticum* pathotypes may be exceptionally similar to each other. One reason for this may be that the development of new *S. endobioticum* pathotypes is caused by very limited changes in avirulence factors rather than extensive genetic recombination between divergent genotypes (Busse et al. 2017). Recently, polymorphic microsatellite markers were used to assess the genetic diversity of potato wart at the intraspecific level for the first time and will certainly contribute to a better understanding of the evolutionary history of this pathogen in the years to come (Gagnon et al. 2016). The molecular methods for *S. endobioticum* detection based on the RealTime PCR (Smith et al. 2014; Bonants et al. 2015) or PNA-based hybridization assay (Duy et al. 2015) have several advantages over the traditional method of microscopic examination in routine diagnostic testing. Molecular methods can significantly reduce the time to disease diagnosis and prevent the spread of *S. endobioticum* to other locations.

Powdery Scab

Alison Lees, Francisco Bittara, and Gary A. Secor

Introduction and Future Perspectives

Spongospora subterranea causes root hyperplasia (root galling) and tuber powdery scab, leading to losses in seed and ware crops of potato worldwide and is also important as the natural vector of Potato mop-top virus (PMTV), an economically important tuber blemish disease of potato found in some regions. Powdery scab is particularly favored by cool, damp conditions and is an intractable disease. Additionally, *S. subterranea* is an unculturable biotroph, making research difficult.

Progress in understanding many aspects of the biology of *S. subterranea* has been slow relative to other plant pathogens and even compared with other plasmodiophorids. However, the development of quantitative molecular assays and the increasing availability of sequence information have allowed recent progress to be made in understanding various aspects of the epidemiology of powdery scab and root galling, and biology of the pathogen.

The pathogen can survive for many years and very low levels of inoculum can cause relatively severe disease outbreaks, making control elusive: it is generally accepted that an integrated approach to disease control will prove most effective as no single control method is totally effective. Additional knowledge of inoculum-based risk, pathogen variation and infection conditions will all contribute towards such a risk-based integrated control system.

The tripartite potato/*S. subterranea* /PMTV interaction is a difficult system to manipulate and there has been little progress in understanding the basis of host specificity and molecular mechanisms of virus transmission. In recent years some progress has been made in developing molecular markers and genomic sequence information for the plasmodiophorids. In addition, host genomics has advanced with the discovery of genetic markers and identification of novel resistance. The availability of in vivo root culture systems for *S. subterranea* propagation, high-throughput nucleotide sequencing, comparative genomics and advanced imaging technologies will no doubt be applied in due course in order to advance knowledge of this intractable and understudied disease and the virus that it vectors. It is envisaged that availability of plasmodiophorid genomes will, in the future, lead to molecular interaction studies between *S. subterranea* and compatible host plants, and will allow the genes or molecular mechanisms involved in host recognition, infection of host cells, multiplication of *S. subterranea* within roots and the development of sporangial or sporogenic phases to be identified.

Causal Organism

The taxonomy of the plasmodiophorids is complicated and has previously been described in detail (Braselton 1995; Dick 2001; Down et al. 2002). The genus *Spongospora*, once ascribed to the fungi, is now considered to be a member of the family *Plasmodiophoridae* within the Super-group Rhizaria, Phylum Cercozoa, under the Class Phytomyxea, which includes members parasitic to higher plants and Stramenopila and which usually cause hypertrophy in the host cells (Bulman and Braselton 2014; Neuhauser et al. 2010). Traditionally, the species *S. subterranea* has been divided into two *formae speciales: Spongospora subterranea* (Wallr.) Lagerheim f.sp. *nasturtii* Tomlinson (the cause of crook root of watercress) and *Spongospora subterranea* (Wallr.) Lagerheim f.sp. *subterranea* (the cause of powdery scab and root galling on potato) (Neuhauser et al. 2010). However, host specificity characteristics, differences in sporangial states and habit as well as molecular data have provided additional evidence to support their placement into the species rank (i.e. *S. subterranea* and *S. nasturtii*; Bulman and Braselton 2014; Dick 2001; Gau et al. 2013; Neuhauser et al. 2010; Qu and Christ 2004) and they will be referred to as such hereafter. Due to the uncertain taxonomy of the group, the collective term "plasmodiophorids" is also commonly used (Braselton 1995). Key features of the plasmodiophorids are that they have zoospores with two anterior undilopodia ("flagella"), multinucleated protoplasts (plasmodia), and environmentally resistant resting spores, and that they are biotrophic parasites. Plasmodiophorids are a monophyletic group with cruciform nuclear division, obligate intracellular parasitism, biflagellated zoospores, and environmentally resistant resting spores (Bulman et al. 2001; Qu and Christ 2004).

A detailed description of life cycle of *S. subterranea* was given by Harrison et al. (1997). There are two major phases in the life cycle of *Spongospora*, each initiated by host cell infection through a single uninucleate plasmodium: in the sporogenic (spore-producing) phase, sporogenic plasmodia are located within infected plant tissue, either in tuber lesions or root, shoot or stolon galls. Following nuclear divisions within the plasmodium, thick-walled resting spores are produced, each being around 3.5–4.5 μm diameter (Jones 1978). These resting spores are aggregated together in sponge-like sporosori which vary in size from 19 to 85 μm. Falloon et al.

(2011) determined the mean numbers of resting spores in sporosori to be about 700. Resting spores may then persist in the soil, where they are able to survive in a dormant state in the absence of potato cultivation for many years, or on tubers. Under suitable conditions, the resting spores germinate and each releases a single biflagellate primary zoospore which can then infect host cells. The mechanisms of host infection are not well understood.

Symptoms

Spongospora subterranea can infect all underground organs of potato (i.e. stolons, tubers, and roots) where the pathogen stimulates the enlargement and division of host cells leading to the appearance of symptoms. Depending on environmental conditions initial tuber symptoms, which take 4–8 weeks to develop, are purplish brown lesions (1–2 mm diameter) that subsequently enlarge into raised mature lesions which burst, exposing large masses of sporosori resulting in characteristic symptoms of the disease.

Sometimes infection of tuber buds can stimulate tubers to swell in that area, forming outgrowths or cankers which may also be infected through lenticels and be covered with scab lesions. Infection of roots or stolons can sometimes lead to development of galls (hyperplasia) , where infected tissue is stimulated to grow and sporosori are formed inside the gall. Galls burst when mature and release the sporosori into the soil. Root or stolon galls can be easily overlooked in the field and there is evidence that varieties with resistance to tuber infection can be susceptible to root or stolon gall production (Falloon 2008), resulting in an unseen build-up of inoculum in the soil.

Fig. Characteristic lesions of powdery scab on cultivar Agria

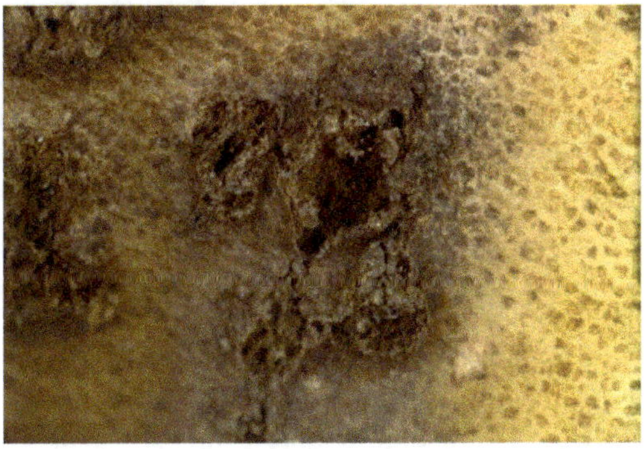

Fig. A typical powdery scab lesion

Fig. Root galling of potato caused by *Spongospora subterranea*

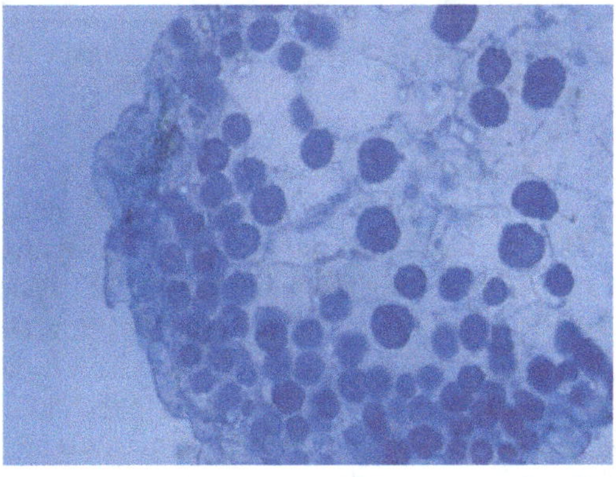

Fig. Sporosori inside root gall

Impact

The main impact of powdery scab is cosmetic, due to a reduction in quality caused by lesions on the tuber surface, thus decreasing their value for either fresh or processing purposes. The disease also affects seed tuber certification, with the extent of the impact dependent on certification tolerances which vary from country to country (Falloon 2008; Wale 2000). Reports of losses due to unmarketable tubers can be as high as 50% in Australia (Hughes 1980) to 100% of the harvested product in Venezuela (Bittara et al. 2009). In addition, root infection by *S. subterranea* can reduce water absorption and nutrient intake (Shah et al. 2012). Some studies have reported a reduction in tuber yield and mean tuber weight due to disease caused by *S. subterranea* on either *Solanum tuberosum* spp. *andigena or S. tuberosum* spp. *tuberosum* (Gilchrist et al. 2011; Shah et al. 2012). In contrast, field studies using the cultivars Shepody and Umatilla Russet (*S. tuberosum* spp. *tuberosum*) showed no detrimental effect on either tuber yield or mean tuber weight due to the development of symptoms on roots and tubers (Johnson and Cummings 2015). *Spongospora subterranea* is also the vector of *Potato mop-top virus* (PMTV), one of the causes of spraing, a tuber blemish disease (Jones and Harrison 1969). PMTV is carried within zoospores and *S. subterranea* can remain viruliferous for many years. Little is known, however, about the virus–vector interactions, particularly with respect to the conditions conducive to transmission of the virus and the differential symptom expression of one or other of the diseases where infection by both organisms has occurred.

Pathogen Populations and Distribution

Since the first description of *Spongospora subterranea* under the name *Erysibe subterranea* by F. W. Wallroth in 1841, the pathogen has been reported in most potato-growing regions of the world, including hot and dry countries where farming is conducted at high altitudes or under irrigation. The number of first reports continues to increase across the world (Harrison et al. 1997; Merz and Falloon 2009; Wale 2000).

Few attempts have been made to characterize variation in *S. subterranea*, an unculturable biotroph with poorly understood genetics. It is difficult to obtain sufficient "clean" DNA for commonly used molecular marker techniques, sequence information is limited and assumptions about what constitutes an isolate or "strain" (sporosorus, resting spore or zoospore) are made. It is not demonstrated that sporosori occur as a result of sexual recombination, although this is thought to be the case (Braselton 1995). In general, analyses have been made on a single sporosori, which may also comprise many genotypes in the form of individual resting spores.

Genetic variation in Internal Transcribed Spacer sequences revealed limited differences among collections from Europe and Australasia and those from South

America (Bulman and Marshall 1998). Three ribotypes (I, II, and III) of *S. subterranea* were identified, with ribotype I, II and III is found in South America, ribotype II in North America and Australasia and ribotypes I and II in Great Britain (Osorio-Giraldo et al. 2012; Qu and Christ 2004). Most recently, Gau et al. (2013) applied SSR markers and ITS and actin sequence data to approximately 700 samples of *S. subterranea* obtained from 19 countries, different resting spore sources of the pathogen (root galls or tuber lesions) and from different potato host (sub) species. They described an overall low genetic diversity in *S. subterranea*, comprising three sample clusters; two occurring in South America (root galls and tuber lesions) and the third group comprising samples from elsewhere, independent of the resting spore source, and suggested that an ecological adaption in the native region due to coevolutionary processes and/or competitive exclusion may have taken place. South American populations were more diverse than those from other regions but no relationship between host species and pathogen diversity was noted.

Host Resistance

Resistance to powdery scab exists in some cultivars worldwide (Torres et al. 1995; Genet et al. 1996; Christ 1987; Falloon et al. 2003; Nitzan et al. 2008; Merz et al. 2012; Bittara et al. 2016). Falloon et al. (2003) found that although cultivars ranged from very susceptible to resistant, all developed zoosporangia and root galls and therefore none possessed immunity. Susceptibility of individual cultivars to root and tuber infection is not always closely correlated (Hughes 1980; Falloon et al. 2003; Bittara et al. 2016), particularly with regards to the relationship between root and tuber infection and root galling (Falloon et al. 2003; van de Graaf et al. 2007). The use of certain cultivars with high tuber resistance may therefore still maintain populations of *S. subterranea* in the soil.

Despite the availability of disease resistant cultivars, genetic resistance to *S. subterranea* currently plays a minor role in disease control as cultivars are usually selected by growers for characteristics other than their ability to resist powdery scab.

Resistance to powdery scab has also been demonstrated in other Solanum species, for example, *S. acaule* was reported to show resistance to pathogen infection in roots (Mäkäräinen et al. 1994) and among potato genotypes reported to have resistance to the formation of root galls at the Columbia Basin (WA) resistant clones were derived from the introgression of *S. bulbocastanum* and the resistant cultivar Summit Russet appeared more than once in their parental background (Nitzan et al. 2008). Although the mechanisms involved in resistance to disease caused by *S. subterranea* are poorly studied, evidence supports the hypothesis that root and tuber resistance to the disease is under control of multiple genes (Bittara et al. 2016; Falloon et al. 2003; Merz et al. 2012). In addition, disease resistance is inherited in an additive manner (Wastie et al. 1988) and is stable across environments (Bittara et al. 2016; Nitzan et al. 2010).

Management

Management of powdery scab is particularly difficult, and no single strategy currently controls powdery scab or root gall formation effectively, making an integrated approach essential (Falloon 2008). Disease avoidance using uncontaminated seed in uninfested soil represents the best method of disease prevention. The relative importance of soil inoculum level in causing disease on tubers was conclusively demonstrated by Brierley et al. (2013) who showed that when arbitrary soil inoculum threshold values of 0, <10 and >10 sporosori/g soil were set, it was observed that the number of crops developing powdery scab increased with the level of inoculum quantified in the field soil preplanting. In field trials carried out to investigate the link between the amount of inoculum added to the soil and disease development, disease incidence and severity on progeny tubers was found to be significantly ($P < 0.01$) greater in plots with increasing levels of inoculum. This information allows disease risk assessments to be made by taking soil inoculum concentration, in combination with other biotic and abiotic factors, into account.

The use of host resistance represents the most effective, sustainable and cost-effective approach for disease management; in addition, genetic resistance is especially suitable on pathogen populations with low genetic diversity (Gau et al. 2013). In a study performed across five European countries, no evidence of genotype x pathogen interaction was observed as disease development as was similar for all locations (Merz et al. 2012). Nevertheless, the use of host resistance is affected by factors defined by the consumer market (Harrison et al. 1997).

Resting spores produced by the pathogen are highly resistant to environmental stresses and can remain viable for >10 years (Merz and Falloon 2009), therefore up to 7 years of crop rotation or more that excludes potatoes might be required to reduce the risk of disease development once the pathogen has stablished in the field (Sparrow et al. 2015). Crop rotation may also have an influence on *S. subterranea* soil infestation as detectable inoculum was shown to be greater following a potato/wheat rotation compared to a potato/pea rotation (Shah et al. 2014). Additionally, a study by Qu and Christ (2006), found four of 16 crop and weed species infected with *S. subterranea* subsequently produced root galls, with sporosori detected on three host species (yellow mustard, oats and tomato). Similarly, Shah et al. (2010) confirmed the presence of sporosori in root galls formed on the solanaceous weeds hairy and black nightshade and subsequently demonstrated pathogenicity on potato with sporosorus inoculum derived from hairy nightshade. It is therefore possible for *S. subterranea* to complete its lifecycle in the absence of potatoes.

It is broadly accepted that high soil water content encourages zoospore release and that soil in which most pore spaces are filled with water facilitates movement of zoospores towards the host, and subsequent infection and disease development. This is supported by associations between seasons of high (or low) rainfall or irrigation treatments with high (or low) levels of powdery scab.

The chemical management of powdery scab is difficult and, in some cases, economically unsuitable. The use of seed and soil-applied chemicals to control powdery scab was previously reviewed in some detail by Merz and Falloon (2009).

Acknowledgements The section on late blight was undertaken as part of, and funded by, the CGIAR Research Program on Roots, Tubers and Bananas (RTB) and supported by CGIAR Trust Fund contributors. The section on wart was founded by the project CORNET (CORNET/2/13/2013) Acronym: SynTest. For the section on powdery scab, the authors thank Dr. Simon Bulman and Professor Richard Falloon, The New Zealand Institute for Plant & food Research for useful discussions regarding the taxonomy of *S. subterranea*. Alison Lees is funded by The Rural and Environment Science and Analytical Services Division of The Scottish Government. Thanks also to Professor Neil Gudmestad, North Dakota State University for helpful discussions.

References

Abuley IK, Nielsen BJ (2017) Evaluation of models to control potato early blight (*Alternaria solani*) in Denmark. Crop Prot 102:118–128

Acuna I, Sagredo B, Gutierrez M, Sandoval C, Fahrenkrog A, Secor G, Rivera V, Mancilla S (2012) Characterization of *Phytophthora infestans* population in Chile. In: Proceedings of the thirteenth EuroBlight workshop, St Petersburg, Russia, 9–12 October 2011. Praktijkonderzoek Plant & Omgeving, PPO, pp 145–150

Ah-Fong AM, Kim KS, Judelson HS (2017) RNA-seq of life stages of the oomycete *Phytophthora infestans* reveals dynamic changes in metabolic, signal transduction, and pathogenesis genes and a major role for calcium signaling in development. BMC Genomics 18(1):198

Andrade-Piedra JL, Cáceres PA, Pumisacho M, Forbes GA (2009) Humans: the neglected corner of the disease tetrahedron—developing a training guide for resource-poor farmers to control potato late blight. Acta Hortic (834):111–122

Anon (2015) CABI/EPPO *Synchytrium endobioticum*. Distribution maps of plant disease, No 10. Wallingford, UK: CABI, Map 1 (Edition 8) http://www.cabi.org/isc/abstract/20153399804

Anon (2017) EPPO Standard PM 7/28 (2) *Synchytrium endobioticum*. Bull OEPP/EPPO Bull 47:420–440

Avenot HF, Sellam A, Karaoglanidis G, Michailides TJ (2008) Characterization of mutations in the iron-sulphur subunit of succinate dehydrogenase correlating with boscalid resistance in *Alternaria alternata* from California pistachio. Phytopathology 98(6):736–742

Avenot H, Sellam A, Michailides T (2009) Characterization of mutations in the membrane-anchored subunits AaSDHC and AaSDHD of succinate dehydrogenase from *Alternaria alternata* isolates conferring field resistance to the fungicide boscalid. Plant Pathol 58(6):1134–1143

Ayad D, Leclerc S, Hamon B, Kedad A, Bouznad Z, Simoneau P (2017) First report of early blight caused by *Alternaria protenta* on potato in Algeria. Plant Dis 101(5):836

Baayen R, Cochius G, Hendriks H, Meffert J, Bakker J, Bekker M, van den Boogert P, Stachewicz H, van Leeuwen G (2006) History of potato wart disease in Europe—a proposal for harmonisation in defining pathotypes. Eur J Plant Pathol 116:21–23

Ballvora A, Flath K, Lübeck J, Strahwald J, Tacke E, Hofferbert H-R, Gebhardt C (2011) Multiple alleles for resistance and susceptibility modulate the defense response in the interaction of tetraploid potato (*Solanum tuberosum*) with *Synchytrium endobioticum* pathotypes 1, 2, 6 and 18. Theor Appl Genet 123:1281–1292

Bartlett DW, Clough JM, Godwin JR, Hall AA, Hamer M, Parr-Dobrzanski B (2002) The strobilurin fungicides. Pest Manag Sci 58:649–652

Bauske MJ, Mallik I, Yellareddygari SKR, Gudmestad NC (2018) Spatial and temporal distribution of mutations conferring QoI and SDHI resistance in *Alternaria solani* across the United States. Plant Dis 102(2):349–358

Bethke PC, Halterman DA, Jansky S (2017) Are we getting better at using wild potato species in light of new tools? Crop Sci 57:1241–1258

Bittara F, Rodriguez D, Sanabria M, Monroy J, Rodriguez J (2009) Fungicides and plant products evaluation for control of powdery scab of potato. Interciencia 34:265–269

Bittara F, Thompson A, Gudmestad N, Secor G (2016) Field evaluation of potato genotypes for resistance to powdery scab on tubers and root gall formation caused by *Spongospora subterranea*. Am J Potato Res 93:497–508

Black W, Mastenbroek C, Mills WR, Peterson LC (1953) A proposal for an international nomenclature of races of *Phytophthora infestans* and of genes controlling immunity in *Solanum demissum* derivatives. Euphytica 2(3):173–179

Boevink P, Wang X, McLellan H, He Q, Naqvi S, Armstrong M, Wei Z, Hein I, Gilroy E, Tian Z, Zhang W, Hein I, Gilroy EM, Tian Z, Birch PRJ (2016) A *Phytophthora infestans* RXLR effector targets plant PP1c isoforms that promote late blight disease. Nat Commun 7:10311

Bonants PJM, van Gent-Pelzer MPE, van Leeuwen GCM, van der Lee TAJ (2015) A real-time TaqMan PCR assay to discriminate between pathotype 1 (D1) and non-pathotype 1 (D1) isolates of *Synchytrium endobioticum*. Eur J Plant Pathol 143:495–506

Bourke A (1993) The visitation of god? The potato and the great Irish famine. Lilliput Press, Ltd., Dublin, Ireland

Braselton J (1995) Current status of the plasmodiophorids. Crit Rev Microbiol 21:263–275

Brierley J, Sullivan L, Wale S, Hilton A, Kiezebrink D, Lees A (2013) Relationship between *Spongospora subterranea* f. sp. *subterranea* soil inoculum level, host resistance and powdery scab on potato tubers in the field. Plant Pathol 62:413–420

Brugmans B, Hutten RGB, Rookmaker N, Visser RGF, van Eck HJ (2006) Exploitation of a marker dense linkage map of potato for positional cloning of a wart disease resistance gene. Theor Appl Genet 112:269–277

Bulman S, Braselton J (2014) Rhizaria: Phytomyxea. In: McLaughlin D, Spatafora J (eds) The Mycota VII, systematics and evolution, Part A, 2nd edn. Springer, Berlin, pp 99–112

Bulman S, Marshall J (1998) Detection of *Spongospora subterranea* in potato tuber lesions using the polymerase chain reaction (PCR). Plant Pathol 47:759–766

Bulman S, Kühn S, Marshal J, Schnepf E (2001) A phylogenetic analysis of the SSU rRNA from members of the Plasmodiophorida and Phagomyxida. Protist 152:43–51

Busse F, Bartkiewicz A, Terefe-Ayana D, Niepold F, Schleusner Y, Flath K, Sommerfeldt-Impe N, Lübeck J, Strahwald J, Tacke E, Hofferbert H-R, Linde M, Przetakiewicz J, Debener T (2017) Genomic and transcriptomic resources for marker development in *Synchytrium endobioticum*, an elusive but severe potato pathogen. Phytopathology 108(3):322–328

Çakır E, Van Leeuwen GCM, Flath K, Meffert JP, Janssen WAP, Maden S (2009) Identification of pathotypes of *Synchytrium endobioticum* found in infested fields in Turkey. EPPO Bull 39(2):175–178

Campo Arana RO, Zambolim L, Costa LC (2007) Potato early blight epidemics and comparison of methods to determine its initial symptoms in a potato field. Rev Fac Nac Agron Medellin 60(2):3877–3890

Cardenas M, Grajales A, Sierra R, Rojas A, Gonzalez-Almario A, Vargas A, Marín M, Fermín G, Lagos LE, Grünwald NJ, Bernal A, Salazar C, Restrepo S (2011) Genetic diversity of *Phytophthora infestans* in the Northern Andean region. BMC Genet 12:23

Casa-Coila VH, Lehner MDS, Hora Júnior BT, Reis A, Nazareno NRX, Mizubuti ES, Gomes CB (2017) First report of *Phytophthora infestans* self-fertile genotypes in Southern Brazil. Plant Dis 101(9):1682–1682

Chowdappa P, Nirmal Kumar BJ, Madhura S, Mohan Kumar SP, Myers KL, Fry WE, Cooke DEL (2015) Severe outbreaks of late blight on potato and tomato in South India caused by recent changes in the *Phytophthora infestans* population. Plant Pathol 64:191–199

Christ BJ (1987) Incidence and severity of powdery scab on 5 potato cultivars. Phytopathology 77(6):985–986

Cooke L, Schepers H, Hermansen A, Bain R, Bradshaw N, Ritchie F, Shaw D, Evenhuis A, Kessel GJT, Wander JGN, Andersson B, Hansen JG, Hannukkala A, Nærstad R, Nielsen BJ (2011) Epidemiology and integrated control of potato late blight in Europe. Potato Res 54(2):183–222

Daayf F, Adam L, Fernando WGD (2003) Comparative screening of bacteria for biological control of potato late blight (strain US-8), using in-vitro, detached-leaves, and whole-plant testing systems. Can J Plant Pathol 25(3):276–284

Danies G, Myers K, Mideros MF, Restrepo S, Martin FN, Cooke DE, Smart CD, Ristaino JB, Seaman AJ, Gugino BK, Grünwald NJ, Fry WE (2014) An ephemeral sexual population of *Phytophthora infestans* in the northeastern United States and Canada. PLoS One 9(12):e116354

Deahl KL, Pagani MC, Vilaro FL, Perez FM, Moravec B, Cooke LR (2003) Characteristics of *Phytophthora infestans* isolates from Uruguay. Eur J Plant Pathol 109(3):277–281

Dick M (2001) Straminipilous fungi: systematics of the Peronosporomycetes including accounts of the marine straminipilous protists, the plasmodiophorids and similar organisms. Kluwer, Dordrecht. 660 p

Domazakis E, Lin X, Aguilera-Galvez C, Wouters D, Bijsterbosch G, Wolters PJ, Vleeshouwers VG (2017) Effectoromics-based identification of cell surface receptors in potato. Methods Mol Biol 2017:337–353

Dong S, Raffaele S, Kamoun S (2015) The two-speed genomes of filamentous pathogens: waltz with plants. Curr Opin Genet Dev 35:57–65

Down G, Grenville L, Clarkson J (2002) Phylogenetic analysis of *Spongospora* and implications for the taxonomic status of the plasmodiophorids. Mycol Res 10:1060–1065

Du J, Verzaux E, Chaparro-Garcia A, Bijsterbosch G, Keizer LP, Zhou J, Liebrand TW, Xie C, Govers F, Robatzek S, Van Der Vossen EA (2015) Elicitin recognition confers enhanced resistance to *Phytophthora infestans* in potato. Nat Plants 1:15034

Duarte HSS, Zambolim L, Rodrigues FA, Paul PA, Pádua JG, Ribeiro Júnior JI, Júnior AFN, Rosado AWC (2014) Field resistance of potato cultivars to foliar early blight and its relationship with foliage maturity and tuber skin types. Trop Plant Pathol 39(4):294–306

Duy J, Smith RL, Collins SD, Connell LB (2015) Rapid colorimetric detection of the fungal phytopathogen *Synchytrium endobioticum* using cyanine dye-indicated PNA hybridization. Am J Potato Res 92:398–409

Fairchild KL, Miles TD, Wharton PS (2013) Assessing fungicide resistance in populations of *Alternaria* in Idaho potato fields. Crop Prot 49:31–39

Fall ML, Tremblay DM, Gobeil-Richard M, Couillard J, Rocheleau H, Van der Heyden H, Lévesque CA, Beaulieu C, Carisse O (2015) Infection efficiency of four *Phytophthora infestans* clonal lineages and DNA Based quantification of sporangia. PLoS One 10(8):e0136312

Falloon R (2008) Control of powdery scab of potato: towards integrated disease management. Am J Potato Res 85:253–260

Falloon R, Genet R, Wallace A, Butler R (2003) Susceptibility of potato (Solanum tuberosum) cultivars to powdery scab (caused by *Spongospora subterranea* f. sp. *subterranea*), and relationships between tuber and root infection. Australas Plant Pathol 32:377–385

Falloon R, Merz U, Lister R, Wallace A, Hayes S (2011) Morphological enumeration of resting spores in sporosori of the plant pathogen *Spongospora subterranea*. Acta Protozool 50:121–132

Flath K, Przetakiewicz J, van Rijswick PCJ, Ristau V, van Leeuwen GCM (2014) Interlaboratory tests for resistance to *Synchytrium endobioticum* in potato by the Glynne-Lemmerzahl method. OEPP/EPPO Bull 44:510–517

Flier WG, Kroon LPNM, Hermansen A, Van Raaij HMG, Speiser B, Tamm L, Fuchs JG, Lambion J, Razzaghian J, Andrivon D, Wilcockson S, Leifert C (2007) Genetic structure and pathogenicity of populations of *Phytophthora infestans* from organic potato crops in France, Norway, Switzerland and the United Kingdom. Plant Pathol 56:562–572

Fonseka DL, Gudmestad NC (2016) Spatial and temporal sensitivity of *Alternaria* species associated with potato foliar diseases to demethylation inhibiting and anilino-pyrimidine fungicides. Plant Dis 100(9):1848–1857

Forbes GA (2012) Using host resistance to manage potato late blight with particular reference to developing countries. Potato Res 55(3–4):205–216

Forbes GA, Escobar XC, Ayala CC, Revelo J, Ordoñez ME, Fry BA, Doucett K, Fry WE (1997) Population genetic structure of *Phytophthora infestans* in Ecuador. Phytopathology 87:375–380

Forbes GA, Goodwin SB, Drenth A, Oyarzun P, Ordoñez ME, Fry WE (1998) A global marker database for Phytophthora infestans. Plant Dis 82:811–818

Forbes GA, Fry WE, Andrade-Piedra JL, Shtienberg D (2008) Simulation models for potato late blight management and ecology. In: Ciancio A, Mukerji K (eds) Integrated management of diseases caused by fungi, phytoplasma and bacteria. Integrated management of plant pests and diseases, vol 3. Springer, Dordrecht

Fry WE (2016) *Phytophthora infestans*: new tools (and old ones) lead to new understanding and precision management. Annu Rev Phytopathol 54:529–547

Fry WE, Birch PRJ, Judelson HS, Grünwald NJ, Danies G, Everts KL, Gevens BK, Gugino BK, Johnson DA, Johnson SB, McGrath MT, Myers KL, Ristaino JB, Roberts PD, Secor G, Smart CD (2015) Five reasons to consider *Phytophthora infestans* a reemerging pathogen. Phytopathology 105(7):966–981

Gagnon MC, vander Lee TAJ, Bonants PJM, Smith DS, Li X, Lévesque CA, Bilodeau GJ (2016) Development of polymorphic microsatellite loci for potato wart from next-generation sequence data. Phytopathology 106:636–644

Gannibal PB, Orina AS, Mironenko NV, Levitin MM (2014) Differentiation of the closely related species, *Alternaria solani* and *A. tomatophila*, by molecular and morphological features and aggressiveness. Eur J Plant Pathol 139:609–623

Garrett KA, Dendy SP (2001) Cultural practices in potato late blight management. In: Complementing resistance to late blight in the Andes, 13-16 February 2001. International Potato Center, Cochabamba, Bolivia

Gau R, Merz U, Falloon R, Brunner P (2013) Global genetics and invasion history of the potato powdery scab pathogen, *Spongospora subterranea* f.sp. *subterranea*. PLoS One 8(6):e67944. https://doi.org/10.1371/journal.pone.0067944

Genet R, Falloon R, Braithwaite M, Wallace A, Fletcher J, Nott H, Braam F (1996) Disease resistance and chemical for control of Powdery scab (*Spongospora subterranea*): progress toward integrated control in New Zealand. In: Abstracts of the 13th triennial conference of the European association for potato research, Veldhoven, The Netherlands, 14–19 July 1996

Ghislain M, Byarugaba AA, Magembe E, Njoroge AW, Rivera C, Roman ML, Tovar JC, Gamboa S, Forbes G, Kreuze JF, Barekye A, Kiggundu A (2018) Stacking three late blight resistance genes from wild species directly into African highland potato varieties confers complete field resistance to local blight races. Plant Biotechnol J. ISSN 1467-7652. Early view 22 Nov 2018. 11 p

Gilchrist E, Soler J, Merz U, Reynaldi S (2011) Powdery scab effect on the potato Solanum tuberosum ssp. andigena growth and yield. Trop Plant Pathol 36:350–355

Goodwin SB, Spielman LJ, Matuszak JM, Bergeron SN, Fry WE (1992) Clonal diversity and genetic differentiation of *Phytophthora infestans* populations in northern and central Mexico. Phytopathology 82:955–961

Gorgiladze L, Meparishvili G, Sikharulidze Z, Natsarishvili K, Meparishvili S (2014) First report of *Synchytrium endobioticum* causing potato wart in Georgia. New Dis Rep 30:4

Goss EM, Tabima JF, Cooke DEL, Restrepo S, Fry WE, Forbes GA, Fieland VJ, Cardenas M, Grünwald NJ (2014) The Irish potato famine pathogen Phytophthora infestans originated in central Mexico rather than the Andes. Proc Natl Acad Sci U S A 111:8791–8796. PMID: 24889615

Groth J, Song Y, Kellermann A, Schwarzfischer A (2013) Molecular characterisation of resistance against potato wart races 1, 2, 6 and 18 in a tetraploid population of potato (*Solanum tuberosum* subsp. *tuberosum*). J Appl Genet 54:169–178

Gupta J (2016) Efficacy of biocontrol agents against *Phytophthora infestans* on potato. Int J Eng Sci Comput 6(9):2249–2251

Haas BJ, Kamoun S, Zody MC, Jiang RHY, Handsaker RE, Cano LM et al (2009) Genome sequence and analysis of the Irish potato famine pathogen *Phytophthora infestans*. Nature 461:393–398

Hampson MC (1993) History, biology and control of potato wart disease in Canada. Can J Plant Pathol 15:223–244

Harbaoui K, Hamada W, Li Y, Vleeshouwers VG, van der Lee T (2014) Increased difficulties to control late blight in Tunisia are caused by a genetically diverse *Phytophthora infestans* population next to the clonal lineage NA-01. Plant Dis 98(7):898–908

Harrison MD, Venette JR (1970) Chemical control of potato early blight and its effect on potato yield. Am J Potato Res 47(3):81–86

Harrison J, Searle R, Williams N (1997) Powdery scab disease of potato—a review. Plant Pathol 46:1–25

Hausladen H, Aselmeyer A (2017) Studies about infection of different *Alternaria solani* isolates on *Solanum tuberosum, Lycopersicon esculentum* and *Solanum nigrum*. PPO-Special Report no 18: 201

Haverkort AJ, Struik PC, Visser RGF, Jacobsen E (2009) Applied biotechnology to combat late blight in potato caused by *Phytophthora infestans*. Potato Res 52(3):249–264

Haverkort AJ, Boonekamp PM, Hutten R, Jacobsen E, Lotz LAP, Kessel GJT, Vossen JH, Visser RGF (2016) Durable late blight resistance in potato through dynamic varieties obtained by cisgenesis: scientific and societal advances in the DuRPh project. Potato Res 59(1):35–66

Hehl R, Faurie E, Hesselbach J, Salamani F, Whitham S, Baker B, Gebhardt C (1999) TMV resistance gene *N* homologues are linked to *Synchytrium endobioticum* resistance in potato. Theor Appl Genet 98:379–386

Hermansen A, Hannukkala A, Hafskjold Nærstad R, Brurberg MB (2000) Variation in populations of *Phytophthora infestans* in Finland and Norway: mating type, metalaxyl resistance and virulence phenotype. Plant Pathol 49:11–22

Horsfield A, Wicks T, Davies K, Wilson D, Paton S (2010) Effect of fungicide use strategies on the control of early blight (*Alternaria solani*) and potato yield. Australas Plant Pathol 39(4):368–375

Hughes I (1980) Powdery scab (*Spongospora subterranea*) of potatoes in Queensland: occurrence, cultivar susceptibility, time of infection, effect of soil pH, chemical control. Aust J Exp Agric Animl Husb 20:625–632

International Potato Center (2007) Annual report 2007: root and tubers the overlooked opportunity. International Potato Center

Jiang RH, Tyler BM (2012) Mechanisms and evolution of virulence in oomycetes. Annu Rev Phytopathol 50:295–318

Jiang R, Li J, Tian Z, Du J, Armstrong M, Baker K, Tze-Yin Lim J, Vossen JH, He H, Portal L, Zhou J, Bonierbale M, Hein I, Lindqvist-Kreuze H, Xie C (2018) Potato late blight field resistance from QTL dPI09c is conferred by the NB-LRR gene R8. J Exp Bot 69(7):1545–1555

Jo KR, Visser RG, Jacobsen E, Vossen JH (2015) Characterization of the late blight resistance in potato differential MaR9 reveals a qualitative resistance gene, R9a, residing in a cluster of Tm-2 2 homologs on chromosome IX. Theor Appl Genet 128(5):931–941

Johanson A, Thurston HD (1990) The effect of cultivar maturity on the resistance of potatoes to early blight caused by *Alternaria solani*. Am J Potato Res 67(9):615–623

Johnson D, Cummings T (2015) Effect of powdery scab root galls on yield of potato in the Columbia Basin. Potato Prog 15:1–12

Jones D (1978) Scanning electron microscopy of cystosori of *Spongospora subterranea*. Trans Br Mycol Soc 70:292–293

Jones R, Harrison B (1969) The behavior pf potato mop-top virus in soil, and evidence for its transmission by *Spongospora subterranea* (Wallr.) Lagerh. Ann Appl Biol 63:1–17

Jones JB, Jones JP, Stall RE, Zitter TA (1993) Compendium of tomato diseases. American Phytopathological Society, St. Paul

Judelson HS (2017) Metabolic diversity and novelties in the oomycetes. Annu Rev Microbiol 71:21–39

Jupe F, Witek K, Verweij W, Śliwka J, Pritchard L, Etherington GJ, Maclean D, Cock PJ, Leggett RM, Bryan GJ, Cardle L (2013) Resistance gene enrichment sequencing (RenSeq) enables reannotation of the NB-LRR gene family from sequenced plant genomes and rapid mapping of resistance loci in segregating populations. Plant J 76(3):530–544

Kamoun S, Furzer O, Jones JD, Judelson HS, Ali GS, Dalio RJ, Roy SG, Schena L, Zambounis A, Panabieres F, Cahill D, Ruocco M, Figueiredo A, Chen XR, Hulvey J, Stam R, Lamour K, Gijzen M, Tyler BM, Grünwald NJ, Mukhtar MS, Tome D, Tor M, Van den Ackerveken G, McDowell J, Daay F, Fry WE, Lindqvist-Kreuze H, Meijer HJ, Petre B, Ristaino J, Yoshida K, Birch PR, Gover F (2014) The top 10 oomycete pathogens in molecular plant pathology. Mol Plant Pathol 16(4):413–434. https://doi.org/10.1111/mpp.12190

Kapsa J (2004) Early blight (*Alternaria spp.*) in potato crops in Poland and results of chemical protection. J Plant Protect Res 44(3):231–238

Kassa B, Sommartya T (2006) Effect of intercropping on potato late blight, *Phytophthora infestans* (Mont.) de Bary development and potato tuber yield in Ethiopia. Kasetsart J (Nat Sci) 40:914–924

Kirk WW, Abu-El Samen FM, Muhinyuza JB, Hammerschmidt R, Douches DS, Thill CA, Groza H, Thompson AL (2005) Evaluation of potato late blight management utilizing host plant resistance and reduced rates and frequencies of fungicide applications. Crop Prot 24(11):961–970

Knaus BJ, Tabima JF, Davis CE, Judelson HS, Grünwald NJ (2016) Genomic analyses of dominant US clonal lineages of *Phytophthora infestans* reveals a shared common ancestry for clonal lineages US11 and US18 and a lack of recently shared ancestry among all other US lineages. Phytopathology 106(11):1393–1403

Kuhn PJ (1984) Mode of action of carboxamides. In: Trinci APJ, Ryley JF (eds) Mode of action of antifungal agents. Cambridge University Press, Cambridge, pp 155–183

Lambert DH, Powelson ML, Stevenson WR (2005) Nutritional interactions influencing diseases of potato. Am J Potato Res 82(4):309–319

Landschoot S, Carrette J, Vandecasteele M, De Baets B, Höfte M, Audenaert K, Haesaert G (2017a) Identification of *A. arborescens*, *A. grandis*, and *A. protenta* as new members of the European *Alternaria* population on potato. Fungal Biol 121:172–188

Landschoot S, Vandecasteele M, Carrette J, De Baets B, Höfte M, Audenaert K, Haesaert G (2017b) Assessing the Belgian potato *Alternaria* population for sensitivity to fungicides with diverse modes of action. Eur J Plant Pathol 148(3):657–672

Larousse M, Galiana E (2017) Microbial partnerships of pathogenic oomycetes. PLoS Pathog 13(1):e1006028. https://doi.org/10.1371/journal.ppat.1006028

Leiminger JH, Hausladen H (2012) Early blight control in potato using disease-orientated threshold values. Plant Dis 96(1):124–130

Leiminger JH, Hausladen H (2014) Untersuchungen zur Befallsentwicklung und Ertragswirkung der Dürrfleckenkrankheit (*Alternaria* spp.) in Kartoffelsorten unterschiedlicher Reifegruppe. Gesunde Pflanzen 66:29–36

Leiminger JH, Auinger H-J, Wenig M, Bahnweg G (2013) Genetic variability among *Alternaria solani* isolates from potatoes in Southern Germany based on RAPD-profiles. J Plant Dis Protect 120(4):164–172

Leiminger JH, Adolf B, Hausladen H (2014) Occurrence of the F129L mutation in *Alternaria solani* populations in Germany in response to QoI application, and its effect on sensitivity. Plant Pathol 63:640–650

Li J, Lindqvist-Kreuze H, Tian Z, Liu J, Song B, Landeo J, Portal L, Gastelo M, Frisancho J, Sanches L, Meijer D, Xie C, Bonierbale M (2012a) Conditional QTL underlying resistance to late blight in a diploid potato population. Theor Appl Genet 124(7):1339–1350

Li Y, van der Lee TAJ, Evenhuis A, van den Bosch GBM, van Bekkum PJ, Förch MG, van Gent-Pelzer MPE, van Raaij HMG, Jacobsen E, Huang SW, Govers F, Vleeshouwers VGAA, Kessel GJT (2012b) Population dynamics of *Phytophthora infestans* in the Netherlands reveals expansion and spread of dominant clonal lineages and virulence in sexual offspring. G3 (Bethesda, Md.) 2(12):1529–1540

Li Y, Shen H, Zhou Q, Qian K, van der Lee T, Huang S (2017) Changing ploidy as a strategy: The Irish Potato Famine pathogen shifts ploidy in relation to its sexuality. Mol Plant-Microbe Interact 30(1):45–52

Liljeroth E, Lankinen Å, Wiik L, Burra DD, Alexandersson E, Andreasson E (2016) Potassium phosphite combined with reduced doses of fungicides provides efficient protection against potato late blight in large-scale field trials. Crop Prot 86:42–55

Lindqvist-Kreuze H, Gastelo M, Perez W, Forbes GA, de Koeyer D, Bonierbale M (2014) Phenotypic stability and genome-wide association study of late blight resistance in potato genotypes adapted to the tropical highlands. Phytopathology 104(6):624–633

Ma Z, Felts D, Michailides TJ (2003) Resistance to azoxystrobin in *Alternaria* isolates from pistachio in California. Pestic Biochem Physiol 77:66–74

MacDonald W, Peters R, Coffin R, Lacroix C (2007) Effect of strobilurin fungicides on control of early blight (*Alternaria solani*) and yield of potatoes grown under two N fertility regimes. Phytoprotection 88(1):9–15

Mäkäräinen E, Rita H, Teperi E, Valkonen J (1994) Resistance to *Spongospora subterranea* in tuber-bearing and non tuber-bearing Solanum spp. Potato Res 37:123–127

Malcolmson JF, Black W (1966) New R genes in *Solanum demissum* Lindl. and their complementary races of *Phytophthora infestans* (Mont.) de Bary. Euphytica 15(2):199–203

Mallik I, Arabiat S, Pasche JS, Bolton MD, Patel JS, Gudmestad NC (2014) Molecular characterization and detection of mutations associated with resistance to succinate dehydrogenase-inhibiting fungicides in *Alternaria solani*. Phytopathology 104(1):40–49

Mariette N, Mabon R, Corbière R, Boulard F, Glais I, Marquer B, Pasco C, Montarry J, Andrivon D (2016) Phenotypic and genotypic changes in French populations of *Phytophthora infestans*: are invasive clones the most aggressive? Plant Pathol 65:577–586

Melnik PA (1998) Wart disease of potato, Synchytrium endobioticum (Schilbersky) Percival. EPPO Technical documents 1032, Paris

Merz U, Falloon R (2009) Review: powdery scab of potato—increased knowledge of pathogen biology and disease epidemiology for effective disease management. Potato Res 52:17–37

Merz U, Lees A, Sullivan L, Schwärzel R, Hebeisen T, Kirk H, Bouchek-Mechiche K, Hofferbert H (2012) Powdery scab resistance in *Solanum tuberosum*: an assessment of cultivar x environment effect. Plant Pathol 61:29–36

Metz N, Adolf B, Chaluppa N, Hückelhoven R, Hausladen H (2019) Occurrence of sdh mutations in German Alternaria solani isolates and potential impact on boscalid sensitivity in vitro, in the greenhouse and in the field. Plant Dis. https://doi.org/10.1094/PDIS-03-19-0617-RE

Montes MS, Nielsen BJ, Schmidt SG, Bødker L, Kjøller R, Rosendahl S (2016) Population genetics of *Phytophthora infestans* in Denmark reveals dominantly clonal populations and specific alleles linked to metalaxyl-M resistance. Plant Pathol 65:744–753

Nærstad R, Hermansen A, Bjor T (2007) Exploiting host resistance to reduce the use of fungicides to control potato late blight. Plant Pathol 56(1):156–166

Nelson R, Orrego R, Ortiz O, Tenorio J, Mundt C, Fredrix M, Vien NV (2001) Working with resource-poor farmers to manage plant diseases. Plant Dis 85:684–695

Neuhauser S, Bulman S, Kirchmair M (2010) Plasmodiophorids: the challenge to understand soil-borne, obligate biotrophs with a multiphasic life cycle. In: Gherbawy Y, Voigt K (eds) Molecular identification of fungi. Springer, Berlin, pp 51–78

Nitzan N, Cummings T, Johnson D, Miller J, Batchelor D, Olsen C, Quick R, Brown C (2008) Resistance to root galling caused by the powdery scab pathogen *Spongospora subterranea* in potato. Plant Dis 92:1643–1649

Nitzan N, Haynes K, Miller J, Johnson D, Cummings T, Batchelor D, Olsen C, Brown C (2010) Genetic stability in potato germplasm for resistance to root galling caused by the pathogen *Spongospora subterranea*. Am J Potato Res 87:497–501

Nowicki M, Foolad MR, Nowakowska M, Kozik EU (2012) Potato and tomato late blight caused by *Phytophthora infestans*: an overview of pathology and resistance breeding. Plant Dis 96:4–17

Obidiegwu JE, Flath K, Gebhardt C (2014) Managing potato wart: a review of present research status and future perspective. Theor Appl Genet 127:763–780

Odilbekov F, Edin E, Garkava-Gustavsson L, Hovmalm HP, Liljeroth E (2016) Genetic diversity and occurrence of the F129L substitutions among isolates of *Alternaria solani* in south-eastern Sweden. Hereditas 153:10

Oh SK, Young C, Lee M, Oliva R, Bozkurt TO, Cano LM, Win J, Bos JIB, Liu HY, Van Damme M, Morgan W, Choi D, Van der Vossen EAG, Vleeshouwers VGAA, Kamoun S (2009) In planta expression screens of *Phytophthora infestans* RXLR effectors reveal diverse phenotypes, including activation of the *Solanum bulbocastanum* disease resistance *protein* Rpi-blb2. Plant Cell 21:2928–2947

Ortiz O, Nelson R, Olanya M, Thiele G, Orrego R, Pradel W, Kakuhenzire R, Woldegiorgis G, Gabriel J, Vallejo J, Xie K (2019) Human and technical dimensions of potato integrated pest management using farmer field schools: International Potato Center and partners' experience with potato late blight management. J Integr Pest Manag 10:1–8

Osorio-Giraldo I, Orozco-Valencia M, Gutiérrez-Sánchez P, González-Jaimes P, Marin-Montoya M (2012) Variabilidad genética de *Spongospora subterranea* f. sp. *subterranea* en Colombia. Bioagro 24:151–162

Pacifico D, Paris R (2016) Effect of organic potato farming on human and environmental health and benefits from new plant breeding techniques. Is it only a matter of public acceptance? Sustainability 8(10):1054

Pacilly FCA, Groot JCJ, Hofstede GJ, Schaap BF, Lammerts van Bueren ET (2016) Analyzing potato late blight control as a social-ecological system using fuzzy cognitive mapping. Agron Sustain Dev 36:35

Pasche JS, Wharam CM, Gudmestad NC (2004) Shift in sensitivity of *Alternaria solani* in response to QoI fungicides. Plant Dis 88(2):181–187

Pasche JS, Piche LM, Gudmestad NC (2005) Effect of the F129L mutation in *Alternaria solani* on fungicides affecting mitochondrial respiration. Plant Dis 89(3):269–278

Perez W, Forbes G (2010) Potato late blight: technical manual. International Potato Center (CIP), Lima. http://www.cipotato.org/publications/pdf/005446.pdf

Perez WG, Gamboa JS, Falcon YV, Coca M, Raymundo RM, Nelson RJ (2001) Genetic structure of Peruvian populations of *Phytophthora infestans*. Phytopathology 91(10):956–965

Phillips SL, Shaw MW, Wolfe MS (2005) The effect of potato variety mixtures on epidemics of late blight in relation to plot size and level of resistance. Ann Appl Biol 147:245–252

Pilet F, Chacón G, Forbes GA, Andrivon D (2006) Protection of susceptible potato cultivars against late blight in mixtures increases with decreasing disease pressure. Phytopathology 96(7):777–783

Plata G (1998) Fenotipos de virulencia en Morochata y tipo sexual de apareamiento en Bolivia de *Phytophthora infestans* que afectan al cultivo de la en la zona de Morochata y determinación del tipo sexual de apareamiento de *Phytophthora infestans* en Bolivia. Tesis Ing. Agr., Facultad de Ciencias Agrícolas Pecuarias y Forestales "Martín Cárdenas" Universidad Mayor de San Simón. Cochabamba, Bolivia

Plich J, Przetakiewicz J, Śliwka J, Flis B, Wasilewicz-Flis I, Zimnoch-Guzowska E (2018) Novel gene Sen2 conferring broad-spectrum resistance to *Synchytrium endobioticum* mapped to potato chromosome XI. Theor Appl Genet 131:2321–2331. In: New locus *Sen2* conferring potato resistance against seven virulent pathotypes of *Synchytrium endobioticum* European Association for Potato Research (EAPR), 20th EAPR Triennial Conference

Przetakiewicz J (2008) Assessment of the resistance of potato cultivars to *Synchytrium endobioticum* (Schilb.) Per. in Poland. EPPO Bull 38:211–215

Przetakiewicz J (2014a) Manual of security sensitive microbes and toxins: *Synchytrium endobioticum*. Liu D (ed). CRC Press, Boca Raton, pp 823–829

Przetakiewicz J (2014b) First report of *Synchytrium endobioticum* (potato wart disease) pathotype 18(T1) in Poland. Plant Dis 98:688

Przetakiewicz J (2015a) First report of new pathotype 39(P1) of *Synchytrium endobioticum* causing potato wart disease in Poland. Plant Dis 99(2):285

Przetakiewicz J (2015b) The viability of winter sporangia of *Synchytrium endobioticum* (Schilb.) Perc. from Poland. Am J Potato Res 92(6):704–708

Przetakiewicz J (2016) A modification of the Potoček's tube test for diagnostic of *Synchytrium endobioticum* (Schilb.) Perc. a causal agent of potato wart disease. Indian Phytopathol 69(4s):260–265

Przetakiewicz J (2017) Sampling, maintenance and pathotype identification of *Synchytrium endobioticum* (schilb.) Perc. Plant Breed Seed Sci 76:29–36

Przetakiewicz J, Plich J (2017) Assessment of potato resistance to *Synchytrium endobioticum*. Plant Breed Seed Sci 76:37–43

Pscheidt JW (1985) Epidemiology and control of potato early blight, caused by *Alternaria solani*. Ph.D. Dissertation. University of Wisconsin, Madison, WI, USA

Qu X, Christ B (2004) Genetic variation and phylogeny of *Spongospora subterranea* f.sp. *subterranean* based on ribosomal DNA sequence analysis. Am J Potato Res 81:385–394

Qu X, Christ B (2006) The host range of *Spongospora subterranea* f.sp. *subterranea* in the United States. Am J Potato Res 83:343–348

Rekad FZ, Cooke DEL, Puglisi I, Randall E, Guenaoui Y, Bouznad Z, Evoli M, Pane A, Schena L, Magnano di San Lio G, Cacciola SO (2017) Characterization of *Phytophthora infestans* populations in northwestern Algeria during 2008-2014. Fungal Biol 121(5):467–477

Rietman H, Bijsterbosch G, Cano LM, Lee HR, Vossen JH, Jacobsen E, Visser RG, Kamoun S, Vleeshouwers VG (2012) Qualitative and quantitative late blight resistance in the potato cultivar Sarpo Mira is determined by the perception of five distinct RXLR effectors. Mol Plant-Microbe Interact 25(7):910–919

Rodenburg SY, Seidl MF, de Ridder D, Govers F (2018) Genome-wide characterization of *Phytophthora infestans* metabolism: a systems biology approach. Mol Plant Pathol 19(6):1403–1413

Rodewald J, Trognitz B (2013) *Solanum* resistance genes against *Phytophthora infestans* and their corresponding avirulence genes. Mol Plant Pathol 14(7):740–757

Rodrigues TTMS, Berbee ML, Simmons EG, Cardoso CR, Reis A, Maffia LA, Mizubuti ESG (2010) First report of *Alternaria tomatophila* and *A. grandis* causing early blight on tomato and potato in Brazil. New Dis Rep 22:28–28

Rojas A, Kirk WW (2016) Phenotypic and genotypic variation in Michigan populations of *Phytophthora infestans* from 2008 to 2010. Plant Pathol 65:1022–1033

Rotem J (1994) The genus *Alternaria*, biology and pathogenicity. APS Press, Minnesota

Runno-Paurson E, Kiiker R, Joutsjoki T, Hannukkala A (2016) High genotypic diversity found among population of *Phytophthora infestans* collected in Estonia. Fungal Biol 120(3):385–392

Saville AC, Martin MD, Ristaino JB (2016) Historic late blight outbreaks caused by a widespread dominant lineage of *Phytophthora infestans* (Mont.) de Bary. PLoS One 11(12):e0168381

Schiek B, Hareau G, Baguma Y, Medakker A, Douches D, Shotkoski F, Ghislain M (2016) Demystification of GM crop costs: releasing late blight resistant potato varieties as public goods in developing countries. Int J Biotechnol 14(2):112–131

Shah F, Falloon R, Bulman S (2010) Nightshade weeds (*Solanum* spp.) confirmed as hosts of the potato pathogens *Meloidogyne fallax* and *Spongospora subterranea* f. sp. *subterranea*. Australas Plant Pathol 39:492–498

Shah F, Falloon R, Butler R, Lister R (2012) Low amounts of *Spongospora subterranea* sporosorus inoculum cause severe powdery scab, root galling and reduced water use in potato (*Solanum tuberosum*). Australas Plant Pathol 41:219–228

Shah F, Falloon R, Butler R, Lister R, Thomas S, Curtin D (2014) Agronomic factors affect powdery scab of potato and amounts of *Spongospora subterranea* DNA in soil. Australas Plant Pathol 43:679–689

Shanthiyaa V, Saravanakumar D, Rajendran L, Karthikeyan G, Prabakar K, Raguchander T (2013) Use of *Chaetomium globosum* for biocontrol of potato late blight disease. Crop Prot 52:33–38

Shtienberg D, Fry WE (1990) Influence of host resistance and crop rotation on initial appearance of potato early blight. Plant Dis 74(11):849–852

Shtienberg D, Blachinsky D, Ben-Hador G, Dinoor A (1996) Effects of growing season and fungicide type on the development of *Alternaria solani* and on potato yield. Plant Dis 80(9):994–998

Smith DC, Rocheleau H, Chapados JT, Abbott C, Ribero S, Redhead SA, Lévesque CA, De Boer SH (2014) Phylogeny of the genus *Synchytrium* and the development of TaqMan PCR assay for sensitive detection of *Synchytrium endobioticum* in soil. Phytopathology 104(4):422–432

Sparrow L, Rettke M, Corkrey S (2015) Eight years of annual monitoring of DNA of soil-borne potato pathogens in farm soils in south eastern Australia. Australas Plant Pathol 44:191–203

Stammler G, Böhme F, Philippi J, Miessner S, Tegge V (2014) Pathogenicity of *Alternaria*-species on potatoes and tomatoes. PPO special report 16:85–96

Torres H, Pacheco M, French E (1995) Resistance of potato to powdery scab (*Spongospora subterranea*) under Andean field conditions. Am Potato J 72:355–363

van de Graaf P, Wale S, Lees A (2007) Factors affecting the incidence and severity of *Spongospora subterranea* infection and galling in potato roots. Plant Pathol 56:1005–1013

van de Vossenberg BTLH, Brankovics B, Nguyen HDT, van Gent-Pelzer MPE, Smith D, Dadej K, Przetakiewicz J, Kreuze J, Boerma M, van Leeuwen G, Levesque CA, van der Lee TAJ (2018a) The linear mitochondrial genome of the quarantine chytrid *Synchytrium endobioticum*; insights into the evolution and recent history of an obligate biotrophic plant pathogen. BMC Evol Biol 18:136

van de Vossenberg BTLH, Westenberg M, Adams I, Afanasenko O, Beniusis A, Besheva A, Boerma M, Bonants P, Choiseul J, Dekker T, Flath K, Heungens K, Karelov A, Przetakiewicz J, Schlenzig A, Yakovleva V, van Leeuwen G (2018b) EUPHRESCO Sendo: an international laboratory comparison study of molecular tests for *Synchytrium endobioticum* detection and identification. Eur J Plant Pathol 151:757–766

Van der Waals JE, Korsten L, Aveling TAS (2001) A review of early blight of potato. Afr Plant Protect 7(2):91–102

Van der Waals JE, Korsten L, Slippers B (2004) Genetic diversity among *Alternaria solani* isolates from potatoes in South Africa. Plant Dis 88:959–964

Venette JR, Harrison MD (1973) Factors affecting infection of potato tubers by *Alternaria solani* in Colorado. Am J Potato Res 50(8):283–292

Vleeshouwers VGAA, Oliver RP (2014) Effectors as tools in disease resistance breeding against biotrophic, hemibiotrophic, and necrotrophic plant pathogens. Mol Plant-Microbe Interact 27:196–206

Vleeshouwers VG, Raffaele S, Vossen JH, Champouret N, Oliva R, Segretin ME, Rietman H, Cano LM, Lokossou A, Kessel G, Pel MA (2011) Understanding and exploiting late blight resistance in the age of effectors. Annu Rev Phytopathol 49:507–531

Volz A, Tongle H, Hausladen H (2013) An integrated concept for early blight control in potatoes. PPO special report 14:12–15

Vossen JH, van Arkel G, Bergervoet M, Jo KR, Jacobsen E, Visser RG (2016) The *Solanum demissum* R8 late blight resistance gene is an Sw-5 homologue that has been deployed worldwide in late blight resistant varieties. Theor Appl Genet 129(9):1785–1796

Wale S (2000) Summary of the session on national potato production and the powdery scab situation. In: Merz U, Lees A (eds) Proc of the 1st European powdery scab workshop. Aberdeen, Scotland, pp 3–9

Wang H, Ren Y, Zhou J, Du J, Hou J, Jiang R, Wang H, Tian Z, Xie C (2017a) The cell death triggered by the nuclear localized RxLR effector PITG_22798 from *Phytophthora infestans* is suppressed by the effector AVR3b. Int J Mol Sci 18(2):409

Wang J, Fernández-Pavía SP, Larsen MM, Garay-Serrano E, Gregorio-Cipriano R, Rodríguez-Alvarado G, Grünwald NJ, Goss EM (2017b) High levels of diversity and population structure in the potato late blight pathogen at the Mexico centre of origin. Mol Ecol 26:1091–1107

Wastie R, Caligari P, Wale (1988) Assessing the resistance of potatoes to powdery scab (*Spongospora subterranea* (Wallr.) Lagerh.). Potato Res 31:167–171

Whisson SC, Boevink PC, Wang S, Birch PRJ (2016) The cell biology of late blight disease. Curr Opin Microbiol 34:127–135

Widmark AK, Andersson B, Cassel-Lundhagen A, Sandström M, Yuen J (2007) *Phytophthora infestans* in a single field in southwest Sweden early in spring: symptoms spatial distribution and genotypic variation. Plant Pathol 56:473–579

Witek K, Jupe F, Witek AI, Baker D, Clark MD, Jones JD (2016) Accelerated cloning of a potato late blight-resistance gene using RenSeq and SMRT sequencing. Nat Biotechnol 34(6):656–660

Woudenberg JHC, Truter M, Groenewald JZ, Crous PW (2014) Large-spored *Alternaria* pathogens in section Porri disentangled. Stud Mycol 79:1–47

Woudenberg JHC, Seidl MF, Groenewald JZ, de Vries M, Stielow JB, Thomma BPHJ, Crous PW (2015) *Alternaria* section *Alternaria*: species, formae speciales or pathotypes? Stud Mycol 82:1–21

Yang L, Wang D, Xu Y, Zhao H, Wang L, Cao X, Chen Y, Chen Q (2017) A new resistance gene against potato late blight originating from *Solanum pinnatisectum* located on potato chromosome 7. Front Plant Sci 8:1729

Yao Y, Li Y, Chen Z, Zheng B, Zhang L, Niu B, Meng J, Li A, Zhang J, Wang Q (2016) Biological control of potato late blight using isolates of *Trichoderma*. Am J Potato Res 93:33–42

Zhu W, Yang LN, Wu EJ, Qin CF, Shang LP, Wang ZH, Zhan J (2015) Limited sexual reproduction and quick turnover in the population genetic structure of *Phytophthora infestans* in Fujian, China. Sci Rep 5:srep10094

Zhu W, Shen L, Fang Z, Yang L, Zhang J, Sun D, Zhan J (2016) Increased frequency of self-fertile isolates in *Phytophthora infestans* may attribute to their higher fitness relative to the A1 isolates. Sci Rep 6, Article number: 29428. https://doi.org/10.1038/srep29428

4

Bacterial Diseases of Potato

Amy Charkowski, Kalpana Sharma, Monica L. Parker, Gary A. Secor, and John Elphinstone

Abstract Bacterial diseases are one of the most important biotic constraints of potato production, especially in tropical and subtropical regions, and in some warm temperate regions of the world. About seven bacterial diseases affect potato worldwide and cause severe damages especially on tubers, the economically most important part of the plant. Bacterial wilt and back leg are considered the most important diseases, whereas potato ring rot, pink eye, and common scab are the minor. Knowledge about zebra chip is extremely rare, as it occurs in a very isolated area and is an emerging disease in New Zealand, Europe, the USA and Mexico. Potato crop losses due to bacterial diseases could be direct and indirect; and they have several dimensions, some with short-term consequences such as yield loss and unmarketability of the produce and others with long-term consequences such as economic, environmental, and social. Some of them are of national and international importance and are the major constraints to clean seed potato production, with considerable indirect effects on trade. This review focuses on *Clavibacter* spp., *Ralstonia* spp., *Pectobacterium* spp., *Dickeya* spp., *Streptomyces* spp., and *Liberibacter* spp. pathogenic to potato, and looks at the respective pathogen in terms of their taxonomy and nomenclature, host range, geographical distribution, symptoms, epidemiology, pathogenicity and resistance, significance and economic losses, and management strategies. Nevertheless, the information collected here deal more with diseases known in developed and developing countries which cause severe economic losses on potato value chain.

A. Charkowski
Colorado State University, Fort Collins, CO, USA
e-mail: amy.charkowski@colostate.edu

K. Sharma (✉) · M. L. Parker
International Potato Center, Nairobi, Kenya
e-mail: kalpanasharma@cgiar.org; m.parker@cgiar.org

G. A. Secor
Department of Plant Pathology, North Dakota State University, Fargo, ND, USA
e-mail: gary.secor@ndsu.edu

J. Elphinstone
Food and Environment Research Agency (Fera) Ltd., Sand Hutton, York, UK
e-mail: john.elphinstone@fera.co.uk

Brown Rot and Bacterial Wilt of Potato Caused by *Ralstonia solanacearum*

Taxonomy and Nomenclature

The genus *Ralstonia* is classified in the ß-Proteobacteria within the family Burkholderiaceae. The species complex of *Ralstonia solanacearum* has long been recognized as a group of phenotypically diverse strains, originally characterized as pathogenic races and biovars (Buddenhagen 1962; Hayward 1964). More recently, Fegan and Prior (2005) described four phylotypes in the species complex, each comprising multiple phylogenetic variants (sequevars) according to sequence diversity within barcoding genes (including 16S rRNA, *hrpB*, *mutS*, and *egl*). Recently, the complex has been reclassified on the basis of whole genome comparisons into three distinct species (Safni et al. 2014; Prior et al. 2016): *R. solanacearum* (Phylotype II), *R. pseudosolanacearum* (Phylotypes I and III), and *R. syzygii* (Phylotype IV).

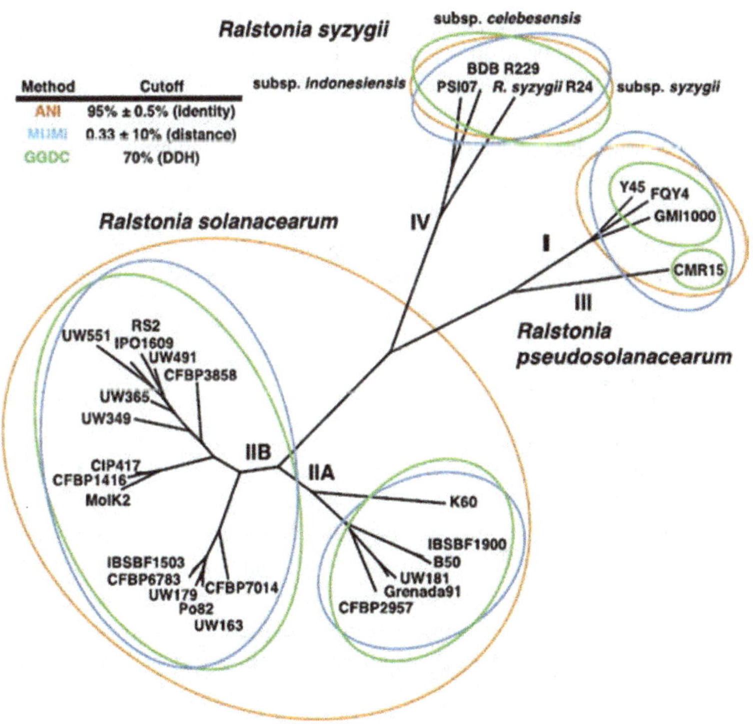

Fig. Phylogenetic relatedness of strains within the *Ralstonia solanacearum* species complex, from Prior et al. (2016). Concatenated tree of distance matrices generated from average nucleotide identity (ANI), maximum unique matches index (MUMi), and genome-to-genome distance calculator (GGDC) to compare DNA:DNA homology (DDH)

Host Range

In addition to potato (*Solanum tuberosum* and *S. phureja*), the large range of economically important hosts includes banana and plantain, cucurbits, eggplant, *Eucalyptus,* ginger, groundnut, mulberry, tobacco, tomato, and many ornamental plants. *R. solanacearum*, *R. pseudosolanacearum*, and *R. syzygii* each comprise strains that were originally designated as race 1 and which occur in tropical areas all over the world, attacking a very wide host range of over 250 hosts in 54 botanical families. Some *R. solanacearum* genotypes within Phylotype II (sequevars IIA-6, IIA-24, IIA-41, IIA-53, IIB-3, IIB-4, and IIB-25), originally described as race 2, cause Moko disease of *Musa* spp. (banana and plantain) and *Heliconia. R. syzygii* comprises three subspecies: subsp. *syzygii* found on clove, subsp. *celebesensis* the cause of banana blood disease and subsp. *indonesiensis* found on solanaceaous crops (potato, tomato, and chilli pepper) as well as clove.

There are strains within each of the four phylotypes that can cause bacterial wilt and brown rot of potato; however, a single strain of *R. solanacearum* (sequevar 1) within Phylotype IIB (PIIB1), formerly known as race 3/biovar 2, is most widely associated with potato. This genotype has a lower *in planta* temperature optimum (27 °C) than most other genotypes (35 °C), often occurring in latent (symptomless) infections at high altitudes in the tropics and in subtropical and temperate potato-growing areas. This strain can also cause bacterial wilt of tomato and can survive in perennial nightshades, which act as secondary hosts. For example, the PIIB1 strain has overwintered in infected underground stolons of *Solanum dulcamara* (woody nightshade), growing along some European rivers, spreading to potato crops when the bacteria were transmitted in contaminated river water which was used for irrigation (Janse et al. 1998). The same strain has also been spread internationally on geranium cuttings produced in Africa and Central America (Williamson et al. 2002).

Geographical Distribution

Phylotype I strains are regarded to be of Asian origin, Phylotype II strains are thought to be of South American origin, whereas Phylotype III appears to have evolved in Africa and Phylotype IV in Indonesia. The *R. solanacearum* species complex is widely designated as a quarantine organism in many countries in an effort to prevent its movement across geographical borders. Nevertheless, the PIIB1 strain has spread from its origin to many potato-growing areas worldwide, presumably with movement in trade of infected seed tubers (Elphinstone 2005). However, this strain has never been reported on potato in the USA, despite it having been introduced on infected geraniums (Williamson et al. 2002) and findings of other Phylotype II strains on potato and other hosts in the southern states. The PIIB1 strain has in fact been designated as a select agent in the USA because of its perceived potential to pose a severe threat to agriculture. Moko disease-causing Phylotype II strains mainly occur in South and Central America and the Carribean, but also appear to have spread to the Philippines where the same strains have been found on cooking banana (ABB and BBB genotypes), causing the so-called bugtok disease.

Symptoms

Wilting is a common symptom of infections of most hosts with all phylotypes. The youngest leaves usually wilt first, appearing at the warmest time of day. Wilting may be visible in only one stem, on one side of a plant or even sectoral in part of a leaf, depending where vascular infections are restricted in sectors of stems and leaf petioles. Leaves may become bronzed or chlorotic and epinasty may occur. Wilting of the whole plant may follow rapidly if environmental conditions are favorable for pathogen growth. As the disease develops, a brown discoloration of the xylem vessels in the stem may be observed above the soil line and adventitious roots may develop. A creamy, slimy mass of bacteria exudes from vascular bundles when the stem is cut. Wilting and collapse of whole plants can lead to rapid death.

Symptoms on infected potato tubers may or may not be visible, depending on the state of development of the disease in relation to the prevailing temperature . Cutting a diseased tuber will reveal browning and necrosis of the vascular ring and in adjacent tissues. A creamy fluid exudate usually appears spontaneously from the vascular ring at the cut surface. Bacterial ooze can emerge from the eyes and stem-end attachment of whole tubers, to which soil adheres. If cut stem or tuber vascular tissue is placed in water, threads of bacterial ooze exude.

Fig. Symptoms of potato brown rot with bacteria oozing from cut vascular tissues (**a**) and eyes (**b**) (UK Crown Copyright—Courtesy of Fera Science Ltd.), and wilted plant in the field (**c**) (Courtesy of International Potato Center)

Epidemiology

Although often described as soilborne pathogens, survival is usually short lived at low temperature in bare soil but is significant in alternative wild host plants (especially perennial nightshade species growing in waterlogged conditions or overwintering volunteers from susceptible crops). The bacteria have been shown to survive in a viable but nonculturable (VBNC) form under stress conditions in soil and water (Kong et al. 2014), but the epidemiological relevance of this is unclear. Disease is usually most severe at temperatures of 24–35 °C, although the PIIB1 strain is more cold tolerant than other strains. High soil moisture or periods of wet weather or rainy seasons are associated with high disease incidence. Entry into plants is usually through root injuries from where the bacteria move by colonization of the xylem where they adhere by polar attraction to the vessel walls, or invade the lumen. Blocking of the vessels by bacterial extracellular polysaccharide (EPS) is considered to be the major cause of wilting. The bacteria can also be transmitted mechanically during pruning operations or when cuttings are taken for propagation.

Long distance movement of vegetative propagating material (e.g. seed potatoes, rhizomes of ginger and turmeric, and banana suckers) can carry latent infections. Natural infection of true seed has only been established for groundnut in Indonesia and China. There have been findings of contaminated seed of other hosts (including tomato, Capsicum, eggplant, and soyabean) although seed infection and transmission has not been substantiated. At present, transmission through water or soil and movement of infected vegetative plant parts are considered to be more important for most host plants than transmission via true seed. In contrast, some strains of *R. solanacearum* and *R. syzygii*, which cause Moko disease and blood disease of banana and Sumatra disease of clove are transmitted by insects (including pollinating flies, bees, wasps, and thrips on banana and xylem-feeding spittlebugs of *Hindola* spp. on clove) with potential for rapid spread over several kilometers.

Pathogenicity Determinants and Resistance

Factors determining pathogenicity, virulence, and avirulence in the bacteria have been recently reviewed (Genin and Denny 2012; Meng 2013). After invasion into root intercellular spaces, expression of *hrpB* is induced, in response to plant signals, activating other *hrp* (hypersensitive response and pathogenicity) genes in construction (*hrpK* and *hrpY*) and regulation (*hrpB* and *hrpG*) of a type 3 secretion system (T3SS), a molecular syringe which is essential for pathogenicity. The bacteria proliferate in intercellular spaces with the aid of a variety of effector proteins, secreted through the T3SS, which suppress plant defenses by interfering with host signal pathways. *HrpB* positively regulates expression not only of *hrp* genes but also of genes encoding a number of plant cell wall-degrading exoproteins secreted through a type II secretion system (T2SS). These include polygalacturonases (PehA, PehB, and PehC), endoglucanase (egl), pectin methylesterase (Pme), and cellobiohydrolase (cbhA), which contribute not only to invasion of xylem vessels, leading to systemic infection, but also to quantitative control of virulence. When cell densities reach a threshold, *PhcA* is activated by build-up of a volatile quorum sensing signal, 3-hydroxy palmitic acid methyl ester (3-OH PAME), inducing genes (*epsABCDEF & P*) controlling biosynthesis of exopolysaccharide (EPS), a major virulence factor. Transcriptome analysis (Jacobs et al. 2012) has also confirmed expression of these and other genes encoding various virulence traits at high cell densities during host infection, including genes imparting stress tolerance (*bcp*, *acrA*, *acrB*, and *dps*) and motility and attachment structures (*pilA* and *fliC*).

Available field resistance to the *R. solanacearum* species complex is limited and tends to be unstable under different environmental conditions and/or strain variability. Traditional breeding has not yet yielded new resistant varieties because of the difficulty in transferring multiple unknown genes from wild germplasm with polygenic resistance into cultivars without cotransferring undesirable linked traits. Furthermore, high-level resistance to host colonization as well as to disease development is needed to avoid the risk of spreading the pathogens in symptomless latent infections. Most studies on the genetic basis of resistance to bacterial wilt have been conducted in the model plants *Arabidopsis thaliana* and *Medicago truncatula* (Huet 2014). Quantitative trait loci (QTL) have been identified that include R genes that encode proteins that recognize bacterial effector avirulence (AVR) proteins, triggering resistance to the bacterium. Transfer of selected R genes from *A. thaliana* into tomato conferred immunity to the *Ralstonia* strain with the corresponding AVR gene (Narusaka et al. 2013). Both broad-spectrum and strain-specific quantitative trait loci (QTLs) have been identified in tomato (Wang et al. 2013), tobacco (Qian et al. 2013) and eggplant (Salgon et al. 2017). Discovery of possible resistance/avirulence (R/Avr) gene for gene resistance mechanisms is particularly interesting since known bacterial effectors can be used to screen for homologous resistance genes in related crops, including potato.

Significance and Economic Loss

Recently ranked by international phytobacteriologists as the second most important of all plant pathogenic bacteria after *Pseudomonas syringae* (Mansfield et al. 2012), the plant pathogenic *Ralstonia* spp. have an extremely wide geographic distribution and host range. On potato alone, it is thought to be responsible for approximately US$1 billion in losses each year, affecting some 3 m farm families over 1.5 m ha in around 80 countries (Elphinstone 2005). Moko disease has affected banana and plantain over thousands of square miles in Central and South America, particularly affecting small susbsistence farmers. In many countries in which the organism has quarantine status, important losses occur as a result of regulatory eradication measures and restrictions introduced on further production on contaminated land.

Management

Disease management remains limited and is hampered by the ability of the pathogens to survive in wet environments on plant debris or in asymptomatic weed hosts, which act as inoculum reservoirs. In the absence of any curative chemical control methods, prevention of bacterial wilt largely relies on the availability of pathogen-free planting material and effective surveillance and monitoring to protect areas free from the bacteria. For potato, effective disease management has mainly resulted from the use of limited generation seed multiplication from pathogen-free nuclear stocks with zero tolerances for the disease in official seed certification programs. Regular post-harvest testing of seed potato tubers is usually also necessary to avoid distribution of latent infections. Similarly, for other vegetatively propagated crops, there is a need to ensure planting material has been tested free of infection and that there are restrictions on the movement of planting material from affected to disease-free areas. Disinfection of pruning and harvesting tools is important in preventing spread of disease e.g. in banana and plantain production. In areas where the pathogen could be spread in contaminated irrigation water, prohibition of irrigation with surface water has been an effective control measure. For hydroponic glasshouse production systems, disinfection of recirculating water (e.g. using chlorine dioxide) can prevent spread of the bacterium. This effectively halted international spread of *R. solanacearum* PIIB1 in geranium cuttings produced in Central America and East Africa following export to the USA and Europe.

Bacterial Blackleg and Tuber Soft Rot Symptoms Caused by *Pectobacterium*

Taxonomy and Nomenclature

The genera *Pectobacterium* is a member of the ß-Proteobacteria in the family *Pectobacteriaceae* within the order *Enterobacterales*. The *Pectobacteriaceae* family also contains the genera *Brenneria*, *Dickeya*, *Lonsdalea*, and *Sodalis* (Adeolu et al. 2016). *Pectobacterium* originally belonged to the genus *Erwinia* (Winslow et al. 1917) with the name *Pectobacterium* being proposed by Waldee (1942). However, the name *Erwinia* persisted until Hauben et al. (1998), using 16S rDNA analysis, re-proposed the name *Pectobacterium*, which has been used since. There are currently 12 species of *Pectobacterium* including *P. aroidearum*, *P. atrosepticum*, *P. betavasculorum*, *P. brasiliense*, *P. cacticida*, *P. carotovorum*, *P. odoriferum*, *P. parmentieri*, *P. peruviense*, *P. polaris*, *P. punjabense*, and *P. wasabiae* (Dees et al. 2017a, b; Khayi et al. 2016; Nabhan et al. 2013; Sarfraz et al. 2018; Waleron et al. 2018; Zhang et al. 2016a, b).

Host Range

Pectobacterium species have a wide range of host plants with approximately a third of these overlapping with hosts for *Dickeya* species (Charkowski 2018; Ma et al. 2007a, b). For *Pectobacterium* species, hosts have been identified in at least 20 dicot families in 13 orders and 12 monocot families in 6 orders, often with only a single isolate being associated with a particular family or order. This may be due to lack of reporting rather than a clear difference in the abilities of these two genera to infect (Charkowski 2018; Ma et al. 2007a, b). However, some important specialization may exist since *Pectobacterium* appears to be found more frequently than *Dickeya* on cabbage, cotton, and mango, and *Dickeya* but not *Pectobacterium* on rice and maize. Some species such as *P. atrosepticum*, *P. betavascularum*, and *P. parmentieri* appear to have a very narrow host range, and *P. aroidearum* appears to be more virulent on monocots than other *Pectobacterium* species (Nabhan et al. 2013). The *Pectobacterium* species most commonly found on potato include *P. atrosepticum*, *P. brasiliense*, *P. carotovorum*, *P. odoriferum*, *P. parmentieri*, *P. peruviense*, *P. polaris*, and *P. punjabense*.

Geographical Distribution

Pectobacterium species are found on all continents where potato is grown, and are likely to be present as saprophytes in the soil, water, and are also regular inhabitants of plant roots when not causing disease. While there are likely to be some regional differences in the species distribution, some appear to be ubiquitous. For example, *P. atrosepticum*, *P. brasiliense*, *P. parmentieri*, and *P. carotovorum* are found on multiple continents (De Boer et al. 2012; Duarte et al. 2004; Kim et al. 2009; Ngadze et al. 2012; Pérombelon and Kelman 1987; Pitman et al. 2008, 2010; She et al. 2017; van der Merwe et al. 2010; van der Wolf et al. 2017; Wang et al. 2017a, b, c).

In Europe, *P. atrosepticum* has been the predominant species responsible for blackleg disease on potato, with *P. carotovorum* responsible a minority of blackleg disease incidents but often associated with soft rot in storage. Recently, at least some of these *P. carotovorum* strains were been reclassified as *P. wasabiae* and subsequently as *P. parmentieri* (Khayi et al. 2016; Nykyri et al. 2012). *P. brasiliense*, which was originally identified as causing disease on potato in Brazil (Duarte et al. 2004), has been common in the United States since at least 2001, as has *P. parmentieri* (Kim et al. 2009; Yap et al. 2004). *P. brasiliense* was not known to cause disease on potato in Europe prior to 2012–2013 but has since increased greatly in its incidence in many European countries (de Werra et al. 2015) and is now recognized as an important pathogen in Africa as well (van der Merwe et al. 2010).

Symptoms

Pectobacterium causes blackleg, which is a stem necrosis that originates from the planted seed tuber (Pérombelon 2002). Necrotic symptoms often extend several centimeters up the stem and necrotic vascular tissue is typically present inside the stem several centimeters beyond where general stem necrosis occurs. The pith of the stem is often decayed. Plant leaves may turn bright yellow and the plant will eventually wilt and die. Infected plants produce few or no tubers. *Pectobacterium* can enter daughter tubers through the xylem or through wounds caused by insects, frost damage, or harvest equipment. Once inside a tuber, it will decay the inside of the tuber, but not the tuber periderm, causing bacterial soft rot. The bacteria will also decay stems damaged by cultivation equipment or severe weather, causing aerial stem rot. In all cases, it is common to find multiple *Pectobacterium* species or *Pectobacterium* and *Dickeya* together when blackleg, aerial stem rot, or soft rot symptoms are present (Kim et al. 2009; Yap et al. 2004). In the United States, *P. partmentieri* is often found with other decay pathogens, such as *Clavibacter michiganensis* and potato rot nematode.

Epidemiology

The most common *Pectobacterium* strains in a region change from year to year and the strains and species present are also not consistent across a particular continent (Dees et al. 2017a, b). The species also differ in optimal and upper limits of growth temperatures. For example, *P. atrosepticum* and *P. parmentieri* die above 33 °C, but *P. carotovorum* and *P. brasiliense* can grow at temperatures up to 39 °C. Initial seed potato production relies on pathogen-free micropropagated plantlets. These plantlets are grown in greenhouses or screen houses to produce minitubers, which are used for field planting. *Pectobacterium* is sometimes found in or on minitubers, but it is more common on potato ones the tubers have been grown in the field. Each generation of potato multiplication tends to increase *Pectobacterium* incidence on potato tubers. Since *Pectobacterium* is common in the environment and can be found in soil, water, weeds, and insects, it is not feasible to produce potatoes free of this pathogen (Charkowski 2015). The bacteria may also be spread by insects (Kloepper and Schroth 1981), but the importance of insects compared to other routes of spread remains unknown. *Pectobacterium* appears to spread mainly at harvest. Bacterial numbers increase dramatically on senescing vines and the bacteria will contaminate harvest equipment and may become aerosolized during harvest. In tubers, the bacteria are found in lenticels and inside the stolon scare. Asymptomatic infestations are common, so it is not possible to visually assess seed potato lots for risk. Blackleg development is highly dependent on the environment and it is unpredictable, even when a seed lot is known to be contaminated with *Pectobacterium*. Detection protocols useful for studying *Pectobacterium* epidemiology were recently complied (Humphris et al. 2015).

Fig. Bacterial blackleg and tuber soft rot symptoms caused by *Pectobacterium* on potato. (**a**) Plants with blackleg are shorter and have curled leaves (A), the stem is blackened on the outside (**b**), the pith inside is decayed and the xylem are brown (**c**). Brown or black decay may spread into leaves (**d**) or leaves may turn bright yellow (**e**). Tubers may have swollen lenticels and sunken lesions (**f**). The soft rot bacteria may enter the tuber through the stolen and decay the center of the tuber (**g**) (Courtesy of Amy O. Charkowski, Colorado State University)

Pathogenicity Determinants and Resistance

Pectobacterium pathogenicity depends upon secreted plant cell wall-degrading enzymes, although several other factors also contribute to virulence (Charkowski et al. 2012). The genetic basis for the observed host range limitation in some *Pectobacterium* species and differences in ability to grow at temperatures above 33 °C remain unknown. There are no examples of gene-for-gene resistance with this necrotrophic pathogen. The antimicrobial peptide Snakin-1 enhances resistance when overexpressed in potato 329–338 and some wild potato species exhibit resistance to *Pectobacterium* (Rietman et al. 2014), but the basis for resistance in wild potato species is poorly understood. There are no resistant commercial potato varieties, but varieties differ in tolerance. There is a large environmental component to disease development for blackleg and little effort has been made to correlate laboratory assays for tolerance with results observed on grower farms. The recent identification of numerous new *Pectobacterium* species suggests that additional novel and potentially high virulent species remain to be discovered and also that this high level of diversity will hinder development of tolerant potato varieties.

Significance and Economic Loss

Pectobacterium has served as a model pathogen for phytobacteriology research for longer than almost any other bacterial pathogen, except *Erwinia amylovora*, and *Pectobacterium* research has resulted in some notable firsts, such as the first demonstration of the role of quorum sensing in bacterial pathogenicity (Pirhonen et al. 1993). It remains an economically significant disease worldwide. Farmers lose millions annually to blackleg, aerial stem rot, and tuber soft rot. Of these, tuber soft rot can be particularly devastating since it occurs after the farmer has invested a full season of inputs into growing the crop.

Management

Pectobacterium management relies mainly on cultural practices (Charkowski 2015; Czajkowski et al. 2011). Growers initiate potato production with micropropagated plantlets that are free of *Pectobacterium*, but tubers are quickly contaminated once they are planted in fields. To reduce the risk of disease at planting, growers should fully suberize seed if they are using cut seed and they should not plant cold seed into wet ground. During the growing season, they should irrigate with ground water if possible and not overfertilize with nitrogen. Rouging infected plants is likely to spread the disease, so this is not recommended. At harvest, the bacteria will multiply on the vines as they senesce, so quickly killing potato vines may aid in reducing disease incidence the following year. Tubers should be allowed to heal before cooling storages. Good airflow and high humidity in potato warehouses will also aid in reducing soft rot in storage. High levels of carbon dioxide in warehouses will promote soft rot development.

Blackleg and Soft Rot of Potato Caused by *Dickeya*

Taxonomy and Nomenclature

The genus *Dickeya* is a member of the ß-Proteobacteria in the family *Pectobacteriaceae* within the order *Enterobacterales*. The *Pectobacteriaceae* family also contains the genera *Brenneria*, *Lonsdalea Pectobacterium*, and *Sodalis* (Adeolu et al. 2016). Members of the *Dickeya* genus originally belonged to the genus *Erwinia* represented by strains within species *E. chrysanthemi* (Burkholder et al. 1953). Later this species was reclassified as *Pectobacterium chrysanthemi* (Hauben et al. 1998), until Samson et al. (2005) elevated the species to the genus *Dickeya* with six species. There have since been some changes and additions to these species, which currently include *D. aquatica*, *D. chrysanthemi*, *D. dadantii*, *D. dianthicola*, *D. fangzhongdai*, *D. paradisiaca*, *D. solani*, and *D. zeae* (Brady et al. 2012; Parkinson et al. 2014; Samson et al. 2005; Tian et al. 2016).

Host Range

Dickeya has a broad host range and can infect plant species in at least 12 dicot families in 10 orders and 10 monocot families in 5 orders, and include ornamentals such as chrysanthemum, carnation, dahlia, and calla lily as well as important crops including carrot, tomato, and, the most economically important, potato (Charkowski 2018; Ma et al. 2007a, b; Samson et al. 2005). While all *Dickeya* species, with the exception of *D. paradisiaca*, have been found on ornamentals in Europe, only *D. dianthicola* and *D. solani* have caused significant economic losses on potato (Toth et al. 2011). In both cases, the lack of genetic diversity between isolates on potato and ornamental hosts suggests the organisms may have spread to potato from such a host (Parkinson et al. 2009; Slawiak et al. 2009). Only *D. aquatica*, which was isolated from waterways in the UK (Parkinson et al. 2014) and Maine (J. Hao, personal communication), has not yet been associated with a plant disease.

Geographical Distribution

As with *Pectobacterium, Dickeya* species have been reported on a wide range of hosts in different countries around the world (Samson et al. 2005). While *D. zeae, D. solani,* and *D. dianthicola* have wide geographic distributions, *D. paradisiaca* appears to be restricted to Colombia (Samson et al. 2005; Toth et al. 2011).

D. dianthicola was the first *Dickeya* species to be associated with plant disease in Europe, occurring on *Dianthus* in the Netherlands, Denmark, and the UK and later spreading to other nations (Hellmers 1958). It was later associated with other ornamentals and crops in a number of European countries, including potato. In some cases, *D. dianthicola* replaced *P. atrosepticum* as the dominant blackleg pathogen (Parkinson et al. 2009; Toth et al. 2011). *D. solani* was recognized independently as a new *Dickeya* pathogen on potato by several groups from 2004 through 2010 (Laurila et al. 2008; Parkinson et al. 2009; Slawiak et al. 2009). Isolates of both *D. dianthicola* and *D. solani* show little genetic diversity compared to isolates from ornamentals, and within these species there is a high degree of genetic similarity. Therefore, it seems likely that these pathogens have independently jumped host from an ornamental onto potato (Toth et al. 2011).

Symptoms

Although *Dickeya* can cause tuber soft rot, it primarily causes blackleg on potato. Blackleg symptoms include necrosis of the potato stem, originating from the mother tuber and spreading several centimeters above ground. Plant leaves will wilt and curl as the disease develops and the plant vascular system will become necrotic. The pith of the stem is often decayed. *D. dianthicola* can also cause severe seed decay and lack of plant emergence in severe cases. Infected plants produce few or no tubers and any tubers produced may decay prior to harvest. Both *Dickeya* and *Pectobacterium* may be present together in diseased plants. In the United States, *P. parmentieri* is the most common species found together with *Dickeya.*

Epidemiology

Initial seed potato production relies on pathogen-free micropropagated plantlets. These plantlets are grown in greenhouses or screenhouses to produce minitubers, which are used for field planting (Frost et al. 2013). *Dickeya* will kill micropropagated plants within a few days and is not typically found in greenhouses or screenhouses. It appears to contaminate potatoes after they have been grown for at least one generation in the field, with the risk of contamination increasing with each

Fig. Foliar symptoms of *Dickeya dianthicola* on potato. Initial symptoms are either a lack of emergence or leaf curling (**a**). The base of the stem turns dark brown or black and this necrosis can extend several centimeters from the soil line (**b**). The pith inside symptomatic stems is often decayed and the xylem are necrotic for several centimeters above the external stem necrosis and the pith decay (**c**). Disease symptoms may only develop on one stem of a multi-stem plant (**d**) (Courtesy of Amy O. Charkowski, Colorado State University)

generation in the field. *Dickeya* does not appear to survive in soil, but it can contaminate waterways and survive for long periods in surface water (Toth et al. 2011). It may also survive in weeds (Fikowicz-Krosko and Czajkowski 2017) or volunteer potatoes and spread by insects (Rossmann et al. 2018). Like *Pectobacterium*, *Dickeya* appears to spread mainly at harvest, where it can spread from infected vines and tubers to previously uncontaminated tubers. The bacteria are mainly found on tuber lenticels, but may also be present in the tuber stolon scar. Asymptomatic infestations are common, so it is not possible to visually assess seed potato lots for risk.

Blackleg development is highly dependent on the environment and it is unpredictable, even when a seed lot is known to be contaminated with *Dickeya*. Plants grown from infested seed lots planted in warm, humid areas tend to develop disease, while plants grown from the same infested seed lot planted in cooler, drier climates may remain healthy. Temperatures above 30 °C during the growing season appear to be particularly conducive to disease development. Co-contamination with *Pectobacterium* and *Dickeya* appears to lead to disease development more frequently than when only *Dickeya* is present.

Pathogenicity Determinants and Resistance

Dickeya pathogenicity relies mainly on pectate lyases and other plant cell wall-degrading enzymes secreted by the bacterial cell, although several other virulence genes are known (Charkowski et al. 2012). Although both *Pectobacterium* and *Dickeya* use plant cell wall-degrading enzymes, there are some important differences in enzyme genes and gene regulation between the genera that may account for some of the differences in disease symptoms. There are no examples of gene-for-gene resistance with *Dickeya* and the basis for resistance to *Dickeya* in wild potato species or for host range is poorly understood. There are no resistant commercial potato varieties, but varieties do differ in tolerance.

Significance and Economic Loss

The relative importance of *Dickeya* as a potato pathogen appears to be increasing (Toth et al. 2011). *D. solani* caused severe losses in the early 2000s in multiple countries and in 2015 *D. dianthicola* was in up to 20% of seed potato lots in some states in the US. Recent development of species-specific PCR assays for *Dickeya* will likely reveal that it is widespread in potato. As with *Pectobacterium*, farmers lose millions annually to blackleg caused by *Dickeya*.

Management

Cultural practices are important for *Dickeya* management and the recommendations are essentially the same as for *Pectobacterium* (Czajkowski et al. 2011, 2013). Growers initiate potato production with micropropagated plantlets that are free of *Dickeya*, but tubers may become contaminated once they are planted in fields. To reduce the risk of disease spread, growers should sanitize equipment thoroughly between seed fields, especially if blackleg is present. At planting, growers should fully suberize seed if they are using cut seed, and they should not plant seed that is too cold or into saturated ground. During the growing season, they should irrigate with ground water if possible and not overfertilize with nitrogen. Rouging infected plants is likely to spread the pathogen if diseased plants are present. At harvest, the *Dickeya* may multiply on the vines as they senesce, so quickly killing potato vines may aid in reducing disease incidence the following year. Tubers should be allowed to heal before cooling storages. Good airflow and high humidity in potato warehouses will also aid in reducing soft rot in storage. High levels of carbon dioxide in warehouses will promote soft rot development. Seed potatoes may be tested for *Dickeya* prior to planting (Czajkowski et al. 2015; Humphris et al. 2015) and growers should avoid planting contaminated seed lots in areas where growing conditions are conducive to blackleg.

Potato Ring Rot Caused by *Clavibacter michiganensis* Subsp. *sepedonicus*

Taxonomy and Nomenclature

Clavibacter michiganensis subsp. *sepedonicus* is a Gram positive, coryneform, aerobic, non-spore-forming bacterium in the Microbacteriaceae family of the Actinobacteria. *C. michiganensis* is the only species currently recognized within the genus; all six of its subspecies (subspp. *Insidiosus*, *michiganensis*, *nebraskensis*, *phaseoli*, *sepedonicus*, and *tesselarius*) are plant pathogens. *C. michiganensis* subsp. *sepedonicus* (*Cms*) was formerly known under the synonyms *Corynebacterium sepedonicum*, *Corynebacterium michiganense* pv. *sepedonicum*, and *Corynebacterium michiganense* subsp. *sepedonicum*.

Host Range

The only economically important host is potato (*Solanum tuberosum*), although natural infection was recently reported for the first time on tomato (van Vaerenbergh et al. 2016). Many members of the Solanaceae, including tomato and eggplant, are susceptible after artifical inoculation. Some solanaceous weeds, e.g. hairy nightshade (*Solanum sarrachoides*) and buffalobur (*S. rostratum*), may harbor the bacterium following potato crops with ring rot (van der Wolf et al. 2005a).

Geographical Distribution

First reported after an outbreak in Germany in 1905 (Appel 1906), it is one of the few major plant pathogens that is not present in the area where the crop evolved, i.e. Andean South America. In North America it was first reported in Quebec (Canada) in 1931 and by 1940 it had spread to all important potato-producing districts in Canada and the USA due to movement in trade of infected seed potato tubers. Subsequent zero tolerances imposed in quarantine and seed certification controls in most potato-growing areas have effectively limited the numbers of findings, although total eradication is difficult. Currently, it tends to occur sporadically in cool, northern latitudes of North America (Northern USA and Canada) with only a single report in Mexico. Strict regulation in Europe has also reduced findings in annual surveys, especially in certified seed crops, with only occasional recent findings in some countries (Bulgaria, Czech Republic, Estonia, Finland, Germany, Greece, Hungary, Latvia, Lithuania, Netherlands, Norway, Slovakia, Sweden, and Turkey). Isolated former outbreaks have been declared eradicated in Austria, Belgium, Cyprus, Denmark, France, Spain, and UK (England and Wales). However, it remains prevalent in areas where formal seed certification is absent (parts of northern, western, and central Russia, Ukraine, Poland, and Romania). It is also reported in Asia (several provinces of China, Japan, Kazakhstan, Korea, Nepal, Pakistan, and Uzbekistan) although its distribution is not clearly defined. Ring rot has never been confirmed in Africa, Australasia, or South America.

In the field, foliar symptoms are not always observed or may occur only at the end of the season when they are difficult to distinguish in the senescing plant and are easily missed during crop inspections. Unlike bacterial wilt, caused by *Ralstonia,* wilting due to the ring rot bacterium is usually slow and initially limited to the leaf margins. Young infected leaves expand more slowly in the infected zones and become distorted. Leaves affected by xylem blockages further down the stem often develop chlorotic, yellow to orange, interveinal areas. Infected leaflets, leaves, and even stems may eventually die. Leaves and tubers may simply be reduced in size and occasionally whole plants can be stunted.

Epidemiology

Factors affecting development and spread of potato ring rot were reviewed in detail by van der Wolf et al. (2005b). Seed potato tubers infected or contaminated with *Cms* are the primary source of infection. Inadvertent dissemination of the bacterium to new places of production occurs with the movement and planting of latently infected seed tubers. The bacterium also spreads from infected tubers through direct contact and via contaminated surfaces of equipment used in potato production, such as seed cutters, planters, harvesters, graders, and transport vehicles as well as in contaminated stores and containers. Plant-to-plant spread in the field is usually low but there is some experimental evidence that insects can transmit the disease (Christie et al. 1991) although the full significance of this is not understood. *Cms* survives for extended periods of many months to years in a dry and cool environment. Its persistence on farm equipment, in stores, and on transport vehicles is an important means by which the bacterium is maintained and spread within farm units and disseminated to other production units.

The bacteria migrate systemically from seed tubers to the stems via the vascular tissue, and subsequently into progeny tubers through the stolons. The pathogen population density increases during the growing season but sometimes can be detected in stems within 3–4 weeks after planting infected seed. Survival of *Cms* in soil is not thought to contribute greatly to ring rot epidemiology although it can overwinter in the field in volunteer tubers (groundkeepers) and in potato tissue

Fig. Symptoms of potato ring rot: initial tuber symptoms (water-soaked vascular ring, bacterial exudate, and start of vascular necrosis) (**a**), necrosis around vascular ring (**b**), advanced necrosis and secondary rotting (**c**), interveinal chlorosis and wilting/epinasty at leaf margin (**d**, **e**), leaf distortion (**f**). (UK Crown Copyright—Courtesy of Fera Science Ltd.)

debris. Survival is longest in cold dry conditions. The bacterium survives particularly well when dried in smears of decayed tuber tissue on equipment, machinery, potato sacks, and storage containers and can remain infectious in the dried state for at least 18 months at temperatures from 5 to −40 °C. *Cms* has been reported to be associated with solanaceous weeds, but any role of these potential inoculum sources in the epidemiology of the ring rot disease of potato is unclear. *Cms* has a low optimum growth temperature (21–23 °C) and is confined mainly to cooler potato-growing regions.

Pathogenicity Determinants and Resistance

The genome of *Cms* was first sequenced by Bentley et al. (2008). In addition to the 3.26 Mb chromosome, which is highly similar amongst all *Clavibacter* subspecies, a 50-kb circular plasmid (pCS1) and a 90-kb linear plasmid (pCSL1) are carried by all *Cms* strains. The plasmid pCS1 is essential for symptom development, but genes required for host recognition, efficient colonization, infection, and evasion or suppression of plant defense are located on the chromosome (Eichenlaub and Gartemann 2011). Two proteins, CelA and Chp-7, have been shown to be required for full virulence (Laine et al. 2000; Nissinen et al. 2001). The gene *celA* is located on the plasmid pCS1 and encodes the cellulase endo-β-1,4-glucanase. Located on the chromosome, *chp-7* encodes a serine protease effector that directly elicits a hypersensitive response in nonhost tobacco plants (Lu et al. 2015). Gene expression studies in *Cms* cells growing in either potato tissue or rich media (Holtsmark et al. 2008) have identified other putative virulence genes. In addition to *celA*, a homologous gene *celB*, two serine proteases and a xylanase was also upregulated in the plant tissue. Three other serine protease genes, including the *chp-7*, were downregulated. Unlike Gram-negative bacterial pathogens of potato, the gram-positive *Clavibacter* pathogens do not have a type 3 secretion system (T3SS) to translocate effectors into host plant cells.

Eichenlaub and Gartemann (2011) have described the likely infection process by *Clavibacter*. During infection the bacteria enter the xylem vessels of host plants, and subsequently spread systemically to colonize the whole plant. *Clavibacter* can be considered as a biotrophic phytopathogen that recruits nutrients (carboxylic acids and sugars) from the xylem fluid. Following growth and colonization, the cell walls of the xylem vessels and surrounding parenchymatic cells are then hydolyzed by expression of cellulases and other extracellular enzymes, potentially including polygalacturonase, pectic lyase, xylanases, and other endoglucanases, leading to symptom development.

There are no currently available potato cultivars with immunity or useful resistance to ring rot. The concept of cultivar tolerance to ring rot is not yet understood and little is known of the status of most commonly grown cultivars with respect to their susceptibility to infection and colonization under varying environmental conditions. Although potato cultivars vary in their propensities to express ring rot symptoms, less variation between cultivars in their susceptibilities to latent infections is observed (De Boer and McCann 1990). Since tolerant cultivars, which tend not to develop symptoms, can act as symptomless carriers of *Cms*, they have been removed from seed certification schemes in North America (Manzer and McKenzie 1988). Laurila et al. (2003) demonstrated that an accession (PI472655) of the wild potato species *Solanum acaule* was susceptible to latent infection by *Cms* at 15 °C but appeared immune to infection at 25 °C.

Significance and Economic Loss

Direct losses due to wilting and tuber rotting in field and store are usually moderate, especially where modern seed certification systems are in place. Nevertheless, ring rot constitutes a constraint on seed potato production, with considerable indirect effects on trade. These result from statutory measures taken against ring rot outbreaks, which include loss of certification, restrictions on further cropping, purchase of new seed stocks, costs of disinfection, and disposal of infected and associated crops and subsequent effects on reputation and export trade.

Management

In the absence of effective chemical or biological control measures, or potato cultivars with adequate levels of resistance, management of potato ring rot must rely on the production and safe distribution of seed potatoes that are free from infection. Control is achieved primarily through strict application of quarantine and seed certification regulations, which involve a zero tolerance for the disease during seed and import inspections and for the pathogen during regular testing of consignments. By laboratory testing for latent infections, infected lots can be detected early and eliminated from seed programs before further spread of the pathogen occurs. Phytosanitary measures must be aimed at the entire potato production system on account of the insidious nature of the disease. Seed potatoes should be imported only from countries which can show, by regular surveys and tests, that they operate a seed-potato production and distribution system free from ring rot.

Implementation of crop rotation, disinfection, and other sanitation practices is most important whenever the disease has occurred to prevent recurrence of the disease and spread of the pathogen. Ring rot infected crops, and any adjacent crops that may have become contaminated should be eliminated from the production system and new certified seed should be acquired for any future production. Disinfectants effective against *Cms* include quaternary ammonia-, chlorine-, or iodine-containing compounds. These should be applied, after cleaning of equipment and other contaminated surfaces, to ensure a minimum of 10 min contact under low organic load. Control of potato volunteers and solanaceaous weeds is also important. The use of whole rather than cut seed helps to reduce any potential spread of the disease.

Common Scab of Potato Caused by *Streptomyces* Species

Taxonomy and Nomenclature

Streptomyces is a gram positive, aerobic, filamentous, spore-forming bacterium in the Streptomycetaceae family of the Actinobacteria. The filamentous mycelia have few or no cross walls. Spores are formed in spiral chains at the tips of hyphae. *Streptomyces* is the largest genus in the Actinobacteria and nearly 600 species are recognized. Most *Streptomyces* are soil-dwelling saprophytes and some species have a beneficial symbiosis with eukaryotes, including plants. At least 12 *Streptomyces* species cause common scab, netted scab, and/or pitted scab on potato. The names of several pathogenic *Streptomyces* species, such as *S. scabies*, were grammatically incorrect when they were first named and the scientific community has only recently begun using corrected names, such as *S. scabiei*.

Common scab is usually caused by *S. scabiei* (Thaxter 1892; Lambert and Loria 1989b), *S. acidiscabiei* (Lambert and Loria 1989a), or *S. turgidiscabiei* (Miyajima et al. 1998). Other species that cause scab symptoms on potato include the pitted scab pathogen *S. caviscabiei* (Goyer et al. 1996), three species first reported in France, including *S. europaeiscabiei*, *S. reticuliscabiei*, and *S. stelliscabiei* (Bouchek-Mechiche et al. 2000), three species first reported in Korea, including *S. luridiscabiei*, *S. niveiscabiei*, and *S. puniciscabiei* (Park et al. 2003), and one species reported in Japan, *S. cheloniumii* (Oniki et al. 1986a, b). The species *S. reticuliscabiei* is genomically the same as *S. turgidiscabiei*, but causes netted scab symptoms rather than typical common scab lesions (Bouchek-Mechiche et al. 2000, 2006). *S. diastatochromogenes* was recently reported as a common scab pathogen of potato, but there is no information available on its relative importance and the species identification was based solely on 16S rDNA sequence (Yang et al. 2017). A related species, *S. ipomeae*, causes root rot of sweet potato. Additional *Streptomyces* species capable of causing common scab have been isolated, but not yet described as species (see Table in Bignell et al. 2014) and additional pathogenic species certainly remain to be discovered. Nonpathogenic strains exist within the pathogenic species, and none of the nonpathogenic strains appear to encode the phytotoxin thaxtomin (Wanner 2006, 2007, 2009).

Host Range

Potato is the most economically important host of plant pathogenic *Streptomyces* species. Plant pathogenic species are also able to cause disease on root crops, such as carrot, beet, parsnip, radish, sweet potato, and turnip (Goyer and Beaulieu 1997), and on peanut pods (Kritzman et al. 1996), but the economic impact of *Streptomyces* on these crops is less important than other diseases that infect these root crops.

Geographical Distribution

Pathogenic *Streptomyces* are present in soils wherever potato is grown and, as the name denotes, the disease it causes is one of the most common and most important potato diseases worldwide. Multiple species are present in individual fields and tubers (Wanner 2009; Lehtonen et al. 2004; Dees et al. 2013). Some species have only been reported from limited geographical regions, but no comprehensive global surveys have been done, so the distribution of pathogenic *Streptomyces* species remains mostly unexplored.

The spread of this pathogen is managed mainly through quality regulations which prohibit planting or shipping of severely affected seed, so there are essentially no limits on the spread of pathogenic *Streptomyces* through seed potatoes. Establishment of new *Streptomyces* strains in field soil is dependent on numerous complex factors, including soil chemistry and resident soil microbes, making establishment of pathogenic *Streptomyces* strains transported on seed potatoes unpredictable.

Symptoms

Streptomyces can cause necrosis on all underground parts of a potato, including roots, stolons, and stems, and it can reduce growth of roots from seed tubers (Han et al. 2008). This pathogen can also cause necrosis on and kill potato seedlings grown from true potato seed. It does not directly cause foliar symptoms, although plant vigor may be reduced due to root necrosis caused by *Streptomyces*.

There is a wide variation in tuber symptoms caused by *Streptomyces*, including pitted scab, erumpent scab, and mild netted scab and symptom type depends, at least in part, on which toxins the infecting strain produces and the potato genotype. The pathogen colonizes tubers as they initiate, often entering the tube through lenticels. Whitish-grey bacterial mycelia and spores are sometimes visible in pitted scab lesions at harvest. The disease does not progress in storage, although tubers with severe pitted scab lesions will dehydrate and will not sprout the following season.

Fig. Common scab symptoms on potato (Courtesy of AHDB Potatoes Sutton Bridge Crop Storage Research)

Epidemiology

Streptomyces has a relatively complex life cycle compared to many bacterial pathogens. It grows vegetatively as filamentous mycelia-like cells. When resources are depleted, the vegetative cells undergo programed cell death, nutrients are transferred to aerial reproductive hyphae, and spores are formed. These hyphae are sometimes visible without magnification inside scab lesions. Pathogenic *Streptomyces* grow best in soils with a pH between 5.2 and 8.0, and temperature of 20–22 °C, which are conditions that also favor potato growth.

Streptomyces survives and disperses mainly through cylindrical spores formed at hyphal tips. The spores can disperse in water, on soil-dwelling invertebrates, and on seed tubers. *Streptomyces* spores can survive in soil for 20 or more years and the spores are heat resistant. The pathogen spores germinate and enter the plant through natural openings, such as lenticels, or through wounds. Tubers are most susceptible to *Streptomyces* colonization during the first month of development. *Streptomyces* cannot cause lesions on mature tubers and lesion size and severity does not progress during storage, although tubers with severe pit scab may become dehydrated and will not sprout the following season.

Because multiple *Streptomyces* species are present in field soil and on diseased plants, epidemiological studies now rely on molecular detection of the species present in order to understand the impacts of management methods, soil characteristics, or biocontrol strains. PCR assays capable of distinguishing *Streptomyces* species are available (Wanner 2009). PCR assays designed to detect genes encoding thaxtomin are also used in epidemiological studies because detection of thaxtomin DNA is correlated with ability of an isolate to cause common scab (Wanner 2006, 2007, 2009; Flores-González et al. 2008) and with development of common scab symptoms in field soils (Qu et al. 2008).

Soils that suppress common scab exist and ongoing work is aimed at identifying the communities that lead to suppressiveness. Soils that suppress common scab have high *Streptomyces* populations. These saprophytic streptomycetes produce antibiotics that inhibit pathogenic *Streptomyces* or that compete with pathogenic *Streptomyces* for resources, thereby reducing common scab (for a comprehensive review, see Schlatter et al. 2017).

Pathogenicity Determinants and Resistance

Bacteria in this genus have unusually large linear genomes of 10–12 Mb and they produce diverse secondary metabolites. In the plant pathogenic *Streptomyces*, large pathogenicity islands encompassing several hundred genes encode virulence genes required for production of secondary metabolites, such as toxins, cytokinin, nitric oxide, and secreted proteins (Bignell et al. 2010; Joshi and Loria 2007). At least two of these pathogenicity islands are mobile (Bukhalid et al. 2002) and one of them can mobilize at least one otherwise nonmobile pathogenicity island (Zhang and Loria 2017). As a result, pathogenicity can be transferred to previously nonpathogenic species (Zhang and Loria 2017).

Phytotoxins are the main *Streptomyces* pathogenicity determinants and the toxin thaxtomin appears to be required for pathogenicity (for a recent review, see Bignell et al. 2014). Thaxtomins, which are nitrated dipeptides (tryptophan and phenylalanine), are required for the development of common scab symptoms (King et al. 1989, 1991; Kinkel et al. 1998). Thaxtomin appears to weaken plant cell walls and cause plant cell hypertrophy through inhibition of cellulose synthesis and cell wall acidification (Fry and Loria 2002; Bischoff et al. 2009). This toxin can be used in potato breeding since seedling tolerance to thaxtomin is correlated with tolerance to common scab in the field (Hiltunen et al. 2011).

The other types of toxins produced by pathogenic *Streptomyces*, including coronatine-like toxins (Fyans et al. 2015), concanamycin (Natsume et al. 2017), borrelidin (Cao et al. 2012), and FD-891 (Natsume et al. 2005), are not necessarily produced by all pathogenic strains and production of these toxins may affect whether an individual strain produces pitted, net, or erumpent common scab symptoms. For example, concanamycin, a type of toxin produced by *S. scabies*, but not by some other *Streptomyces* species, may be required for formation of pitted scab lesions and appears to be synergistic with thaxtomin (Natsume et al. 2017).

Enzymes may also play a role in *Streptomyces* pathogenicity. *Streptomyces* lesions typically do not autofluoresce, suggesting that suberin formation is either inhibited or digested. Two genes that encode potential suberinases are present in the *S. scabies* genome and biochemical evidence supports that suberin is degraded (Beaulieu et al. 2016; Komeil et al. 2013). Degradation of suberin also appears to increase expression of the numerous cellulases produced by *S. scabies* (Padilla-Reynaud et al. 2015). *Streptomyces* toxin production is induced by plant-derived molecules, including the disaccharide cellobiose, a breakdown product of cellulose.

Little is known about the genetic basis of resistance to common scab. Suggested mechanisms include phellum layer thickness (Thangavel et al. 2016), phellum suberization (Thangavel et al. 2016; Khatri et al. 2011), detoxification of thaxtomin (Acuna et al. 2001), or sustained expression of disease defense genes (Merete Wiken Dees et al. 2016). Differences in ability of potato varieties to support growth of nonpathogenic *Streptomyces* species may also affect susceptibility to common scab (Wanner 2007).

Significance and Economic Loss

Common scab can cause complete loss, although this is usually associated with mismanagement of the crop, such as adding too much lime to a field, insufficient irrigation, or highly susceptible varieties planted in fields with high disease pressure.

Direct losses occur annually, however, worldwide, and common scab is often listed among the most important potato diseases (for example, Hill and Lazarovits 2005).

Management

The best option is disease tolerance or resistance, but currently there are limited options for potato varieties with high tolerance to common scab. Common scab symptom development is affected by soil moisture and chemistry, the soil microbial community, crop rotation, and host genetics in a complex manner that has made predicting common scab severity and managing this disease difficult. A comprehensive review of these challenges was published by Dees and Wanner (2012). Recommendations for management of common scab usually include adequate irrigation during tuber formation, and low soil pH (<5.2). Typically, sulfur fertilizers are used to reduce soil pH and this can reduce disease severity (Pavlista 2005). However, these methods sometimes fail to provide adequate management and can lead to other production problems. For example, over-irrigation during tuber formation can lead to development of powdery scab and several other potato diseases, and low soil pH limits farmer options for crop rotations and selects for *S. acidiscabies*.

Chemical treatments can work for a season, but are often expensive and damaging to the soil, making this the least sustainable disease management option. Some commonly used effective chemicals include fludioxonil as a seed piece treatment, chloropicrin as a soil fumigant, and pentachloronitrobenzene as an in-furrow treatment (Al-Mughrabi et al. 2016; Powelson and Rowe 2008). Fluazinam may also provide some control of common scab (Santos-Cervantes et al. 2017).

Crop rotation choices can also reduce common scab severity (Powelson and Rowe 2008; Larkin and Halloran 2014; Larkin et al. 2011; Larkin and Griffin 2007). These crop rotations tend to include brassica crops as a biofumigant and commonly planted green manures that are allelopathic and that help control multiple soil-borne potato diseases. Soil amendments, such as rice bran, chelated iron, or peat can decrease common scab, likely by increasing the population of nonpathogenic streptomycetes (Tomihama et al. 2016; Sarikhani et al. 2017). Some soil amendments, such as manure, which increases soil pH, will increase common scab severity.

Biocontrol with nonpathogenic *Streptomyces* strains also shows promise and the mechanism of biocontrol is likely similar to that seen in suppressive soils, which is thought to be due to both resource completion and antibiotic production (Schlatter et al. 2017). Suppressive soils develop through repeated monoculture of potato, but this practice results in accumulation of other soil-borne pathogens. However, repeated inoculations of soils with a single antagonistic *Streptomyces* strain can result in common scab suppression in as little as 3 years, and suppressive lasted for 2 years beyond the last inoculation (Hiltunen et al. 2017).

Zebra Chip of Potato Caused by *Liberibacter*

Taxonomy and Nomenclature

The genus "*Candidatus* Liberibacter" is a gram-negative bacterium in the Rhizobeaceae family. At least seven *Ca. Liberibacter* species exist. Of these, "*Candidatus* Liberibacter solanacearum" (Lso), which is a phloem-limited pathogen, is the only known potato-infecting species. There are at least five Lso haplotypes, with haplotypes A and B causing disease on potato and the remaining three haplotypes infecting carrots and celery (Nelson et al. 2011; Teresani et al. 2014). At 1.26 Mbp, the circular Lso genome is relatively small (Hong Lin et al. 2011) and there are relatively few genomic differences among Lso haplotypes (Wang et al. 2017a, b, c). Compared to related free-living bacteria, such as *Agrobacterium*, Lso has a low G + C content and lacks many genes involved in metabolism.

Host Range

Potato is the most economically important host of Lso haplotypes A and B, but Lso can also infect other solanaceous crops and weeds. All Ca. Liberibacter are spread by *Bactericera* species and Lso also infects its vector and can reduce vector fitness (Yao et al. 2016). Although Lso is only spread in potato by *B. cockerelli*, but it can also be found in other *Bactericera* species, suggesting that vector feeding preferences limit the species of vectors important for zebra chip and not Lso-vector interactions (Borges et al. 2017).

Geographical Distribution

Potato psyllids are native to North and Central America, and it recently invaded New Zealand (Teulon et al. 2009). The bacterial pathogen has spread with its vector and can be found wherever potato psyllids are found. The highest disease incidence is typically found in the Southwestern United States, Mexico, and Central America.

Symptoms

Zebra chips symptoms are severe on both the foliage and the tubers. The upper parts of infected plants have leaf curling, chlorosis, shortened internodes, aerial tubers, and early necrosis and death (Buchman et al. 2012). The tubers appear to have glassy or brown streaks that darken when they are fried, giving the

Fig. Typical Lso symptoms on potato. Leaf curl symptoms on foliage (**a**); Necrotic specks in tuber (**b**); Blackened splotches on a fried chip (healthy on left, and diseased on right) (**d**) (Courtesy of Gary A. Secor, North Dakota State University)

disease its name, zebra chip. Tuber development slows or ceases in symptomatic plants, resulting in yield losses. Lso appears to reduce protease inhibitor levels in tubers, and as a result, tubers from infected plants have less protein (Kumar et al. 2015). Infected tubers either do not sprout or have only hair sprouts (Rashed et al. 2015). If plants emerge from infected tubers, they die shortly after emergence. Storage temperature affects symptom development, with cooler storage (3 °C) resulting more tuber symptoms than warmer storage temperatures (6 or 9 °C) (Wallis et al. 2017).

Epidemiology

Lso is transmitted solely by *B. cockerelli*, which feeds on phloem with its piercing-sucking mouthparts. The pathogen is transmitted in a persistent, propagative, and circulative fashion and it is also transmitted transovarially in the psyllid (Cicero et al. 2016; Hansen et al. 2008). A 2-week latent period occurs between psyllid acquisition of Lso and ability to transmit the pathogen (Sengoda et al. 2014). In most regions in North America where the pathogen and vector are present, potatoes are infected late in the season. Surprisingly, Lso reduces the fitness of its insect vector, with haplotype B resulting in more insect mortality than haplotype A (Yao et al. 2016).

Once transmitted into a leaf, the pathogen does not cause symptoms for at least 3 weeks. It is also not evenly distributed in plants, which makes it difficult to detect prior to symptom development and this has hampered epidemiological studies. Lso is not culturable, which also makes epidemiological studies more challenging and as a result, researchers rely mainly on PCR assays for pathogen detection (Ananthakrishnan et al. 2013; Secor et al. 2009). Based on symptom development, Lso appears to be sensitive to temperatures above 32 °C and to thrive at 27–32 °C. Solanaceous weeds serve as important reservoirs for Lso and can provide a green bridge between potato crops (Thinakaran et al. 2015).

Infected tubers rarely sprout and when they do, they tend to develop hair sprouts. As a result, this disease is poorly transmitted through seed potatoes and insect transmission remains the most important mode of spread. For this reason, zebra chip is not currently regulated through seed potato certification in North America. It has, however, impacted export of potatoes from North America.

Pathogenicity Determinants and Resistance

Liberibacter pathogenicity determinants were recently thoroughly reviewed (Wang et al. 2017a, b, c). There are no resistant potato varieties, although timing and severity of symptoms differ among varieties (Lévy et al. 2015). Tolerant lines still support Lso levels similar to those found in susceptible varieties, but the Lso has less impact on plant physiology and symptom development in tubers in tolerant lines (Rashidi et al. 2017; Wallis et al. 2015). Recent results also suggest that psyllids are not able to transmit the pathogen with equal efficiency into all potato lines (Rashidi et al. 2017).

Significance and Economic Loss

Zebra chip has caused millions in losses in North America, and although seed tubers are not a major source of inoculum, it has affected the potato export market. The spread of this pathogen and its vector to New Zealand has also caused significant losses there. In addition to losses in yield and quality, the high cost of vector management has added to financial losses caused by *L. solanacearum*.

Management

Insecticides are the main management method used for control of zebra chip. Growers in North America monitor psyllids and determine when psyllids appear and the percentage of Lso-infected psyllids present. They may spray insecticides a dozen or more times during the growing season to protect the potato crop, with imidocloprid and spirotetramat among the most commonly used (Guenthner et al.

2012). Since the insects tend to be present on the underside of leaves, effectively covering the underside of the leaves is essential. These sprays are expensive and the potential for insecticide resistance and loss of natural enemies due to frequent sprays makes this approach unsustainable in the long term.

Concluding Remarks

Bacterial diseases of potato have remained an economically significant disease worldwide. Farmers lose millions of dollars annually due to bacterial diseases. Bacterial wilt, soft rot, and ring rot have got international attention as they constitute a huge constraint on seed potato production, with considerable indirect effects on trade. Rigorous seed certification and testing programs in developed countries have limited the impact of these diseases within their value chains, while developing countries commonly lack these safeguards. Lack of certified disease-free planting material in many developing country contexts contributes to further distribution of these pathogens via latently infected tubers, as well as tuber quality and yield degeneration caused by farmers replanting diseased seed year to year. Bacterial disease management efforts in developing countries should follow the systems approach that incorporates specific operational practices to reduce the likelihood of incursion, establishment, and growth of these pathogens in potato crops. This includes training farmers in proper production practices, on-farm management tools, using healthy seed tubers, and planting in clean soils. Additional factors to consider in controlling these diseases can include sanitation, cultural practices, crop rotation with nonhost plants, and the use of tolerant or resistant varieties.

References

Acuna IA, Strobel GA, Jacobsen BJ, Corsini DL (2001) Glucosylation as a mechanism of resistance to thaxtomin A in potatoes. Plant Sci 161(1):77–88

Adeolu M, Alnajar S, Naushad S, Gupta RS (2016) Genome-based phylogeny and taxonomy of the ‘Enterobacteriales’: proposal for Enterobacterales ord. nov. divided into the families Enterobacteriaceae, Erwiniaceae fam. nov., Pectobacteriaceae fam. nov., Yersiniaceae fam. nov., Hafniaceae fam. nov., Morganellaceae fam. nov., and Budviciaceae fam. nov. Int J Syst Evol Microbiol 66(12):5575–5599. https://doi.org/10.1099/ijsem.0.001485

Al-Mughrabi KI, Vikram A, Poirier R, Jayasuriya K, Moreau G (2016) Management of common scab of potato in the field using biopesticides, fungicides, soil additives, or soil fumigants. Biocontrol Sci Tech 26(1):125–135

Ananthakrishnan G, Choudhary N, Roy A, Sengoda VG, Postnikova E, Hartung JS, Stone AL, Damsteegt VD, Schneider WL, Munyaneza JE (2013) Development of primers and probes for genus and species specific detection of ‘Candidatus Liberibacter species’ by real-time PCR. Plant Dis 97(9):1235–1243

Appel O (1906) Neuere Untersuchungen Über Kartoffel-Und Tomaten-Erkrankungen. Borntraeger, Stuttgart

Beaulieu C, Sidibé A, Jabloune R, Simao-Beaunoir A-M, Lerat S, Monga E, Bernards MA (2016) Physical, chemical and proteomic evidence of potato suberin degradation by the plant pathogenic bacterium Streptomyces scabiei. Microbes Environ 31(4):427–434

Bentley SD, Corton C, Brown SE, Barron A, Clark L, Doggett J, Harris B, Ormond D, Quail MA, May G (2008) Genome of the actinomycete plant pathogen Clavibacter michiganensis subsp. sepedonicus suggests recent niche adaptation. J Bacteriol 190(6):2150–2160

Bignell DRD, Huguet-Tapia JC, Joshi MV, Pettis GS, Loria R (2010) What does it take to be a plant pathogen: genomic insights from Streptomyces species. Antonie Van Leeuwenhoek 98(2):179–194

Bignell DRD, Fyans JK, Cheng Z (2014) Phytotoxins produced by plant pathogenic Streptomyces species. J Appl Microbiol 116(2):223–235

Bischoff V, Cookson SJ, Wu S, Scheible W-R (2009) Thaxtomin A affects CESA-complex density, expression of cell wall genes, cell wall composition, and causes ectopic lignification in Arabidopsis thaliana seedlings. J Exp Bot 60(3):955–965

Borges KM, Rodney Cooper W, Garczynski SF, Thinakaran J, Jensen AS, Horton DR, Munyaneza JE, Cueva I, Barcenas NM (2017) 'Candidatus Liberibacter solanacearum' associated with the psyllid, Bactericera maculipennis (Hemiptera: Triozidae). Environ Entomol 46(2):210–216

Bouchek-Mechiche K, Gardan L, Normand P, Jouan B (2000) DNA relatedness among strains of Streptomyces pathogenic to potato in France: description of three new species, S. europaeiscabiei sp. nov. and S. stelliscabiei sp. nov. associated with common scab, and S. reticuliscabiei sp. nov. associated with netted scab. Int J Syst Evol Microbiol 50(1):91–99

Bouchek-Mechiche K, Gardan L, Andrivon D, Normand P (2006) Streptomyces turgidiscabies and Streptomyces reticuliscabiei: one genomic species, two pathogenic groups. Int J Syst Evol Microbiol 56(12):2771–2776

Brady CL, Cleenwerck I, Denman S, Venter SN, Rodríguez-Palenzuela P, Coutinho TA, De Vos P (2012) Proposal to reclassify Brenneria quercina (Hildebrand and Schroth 1967) Hauben et Al. 1999 into a new genus, Lonsdalea gen. nov., as Lonsdalea quercina comb. nov., descriptions of Lonsdalea quercina subsp. quercina comb. nov., Lonsdalea quercina subsp. iberica subsp. nov. and Lonsdalea quercina subsp. britannica subsp. nov., emendation of the description of the genus Brenneria, reclassification of Dickeya dieffenbachiae as Dickeya dadantii subsp. Dieffenbachiae comb. nov., and emendation of the description of Dickeya dadantii. Int J Syst Evol Microbiol 62(7):1592–1602. https://doi.org/10.1099/ijs.0.035055-0

Buchman JL, Fisher TW, Sengoda VG, Munyaneza JE (2012) Zebra chip progression: from inoculation of potato plants with Liberibacter to development of disease symptoms in tubers. Am J Potato Res 89(2):159–168

Buddenhagen IW (1962) Designations of races in pseudomonas Solanacearum. Phytopathology 52.726

Bukhalid RA, Takeuchi T, Labeda D, Loria R (2002) Horizontal transfer of the plant virulence gene, Nec1, and flanking sequences among genetically distinct Streptomyces strains in the Diastatochromogenes cluster. Appl Environ Microbiol 68(2):738–744

Burkholder WR, Mcfadden LA, Dimock EW (1953) A bacterial blight of chrysanthemums. Phytopathology 43(9). https://www.cabdirect.org/cabdirect/abstract/19541101119

Cao Z, Khodakaramian G, Arakawa K, Kinashi H (2012) Isolation of Borrelidin as a phytotoxic compound from a potato pathogenic Streptomyces strain. Biosci Biotechnol Biochem 76(2):353–357

Charkowski AO (2015) Biology and control of Pectobacterium in potato. Am J Potato Res 92(2):223–229

Charkowski AO (2018) The changing face of bacterial soft-rot diseases. Annu Rev Phytopathol 56(1):269–288. https://doi.org/10.1146/annurev-phyto-080417-045906

Charkowski A, Blanco C, Condemine G, Expert D, Franza T, Hayes C, Hugouvieux-Cotte-Pattat N et al (2012) The role of secretion systems and small molecules in soft-rot Enterobacteriaceae pathogenicity. Annu Rev Phytopathol 50(1):425–449. https://doi.org/10.1146/annurev-phyto-081211-173013

Christie RD, Sumalde AC, Schulz JT, Gudmestad NC (1991) Insect transmission of the bacterial ring rot pathogen. Am Potato J 68(6):363–372

Cicero JM, Fisher TW, Qureshi JA, Stansly PA, Brown JK (2016) Colonization and intrusive invasion of potato Psyllid by 'Candidatus Liberibacter solanacearum'. Phytopathology 107(1):36–49

Czajkowski R, Pérombelon MCM, van Veen JA, van der Wolf JM (2011) Control of blackleg and tuber soft rot of potato caused by Pectobacterium and Dickeya species: a review. Plant Pathol 60(6):999–1013. https://doi.org/10.1111/j.1365-3059.2011.02470.x

Czajkowski R, de Boer WJ, van der Wolf JM (2013) Chemical disinfectants can reduce potato blackleg caused by 'Dickeya solani'. Eur J Plant Pathol 136(2):419–432. https://doi.org/10.1007/s10658-013-0177-8

Czajkowski R, Pérombelon MCM, Jafra S, Lojkowska E, Potrykus M, van der Wolf JM, Sledz W (2015) Detection, identification and differentiation of Pectobacterium and Dickeya species causing potato blackleg and tuber soft rot: a review. Ann Appl Biol 166(1):18–38. https://doi.org/10.1111/aab.12166

De Boer SH, McCann M (1990) Detection of Corynebacterium sepedonicum in potato cultivars with different propensities to express ring rot symptoms. Am Potato J 67(10):685

De Boer SH, Li X, Ward LJ (2012) Pectobacterium spp. associated with bacterial stem rot syndrome of potato in Canada. Phytopathology 102(10):937–947

de Werra P, Bussereau F, Keiser A, Ziegler D (2015) First report of potato blackleg caused by Pectobacterium carotovorum subsp. Brasiliense in Switzerland. Plant Dis 99(4):551

Dees MW, Wanner LA (2012) In search of better management of potato common scab. Potato Res 55(3–4):249–268

Dees MW, Sletten A, Hermansen A (2013) Isolation and characterization of Streptomyces species from potato common scab lesions in Norway. Plant Pathol 62(1):217–225

Dees MW, Lysøe E, Alsheikh M, Davik J, Brurberg MB (2016) Resistance to Streptomyces turgidiscabies in potato involves an early and sustained transcriptional reprogramming at initial stages of tuber formation. Mol Plant Pathol 17(5):703–713

Dees MW, Lebecka R, Perminow JIS, Czajkowski R, Motyka A, Zoledowska S, Śliwka J, Lojkowska E, Brurberg MB (2017a) Characterization of Dickeya and Pectobacterium strains obtained from diseased potato plants in different climatic conditions of Norway and Poland. Eur J Plant Pathol 148(4):839–851

Dees MW, Lysøe E, Rossmann S, Perminow J, Brurberg MB (2017b) Pectobacterium polaris sp. nov., isolated from potato (Solanum tuberosum). Int J Syst Evol Microbiol 67(12):5222–5229

Duarte V, De Boer SH, Ward LJ, De Oliveira AMR (2004) Characterization of atypical Erwinia carotovora strains causing blackleg of potato in Brazil. J Appl Microbiol 96(3):535–545

Eichenlaub R, Gartemann K-H (2011) The Clavibacter michiganensis subspecies: molecular investigation of gram-positive bacterial plant pathogens. Annu Rev Phytopathol 49:445–464

Elphinstone JG (2005) The current bacterial wilt situation: a global overview. In: Allen C, Piror P, Hayward AC (eds) Bacterial wilt disease and the Ralstonia solanacearum species complex. APS Press, St Paul, MN, pp 9–28

Fegan M, Prior P (2005) How complex is the Ralstonia solanacearum species complex. APS Press, St. Paul

Fikowicz-Krosko J, Czajkowski R (2017) Systemic colonization and expression of disease symptoms on bittersweet nightshade (Solanum dulcamara) infected with a GFP-tagged Dickeya solani IPO2222 (IPO2254). Plant Dis 102(3):619–627. https://doi.org/10.1094/PDIS-08-17-1147-RE

Flores-González R, Velasco I, Montes F (2008) Detection and characterization of Streptomyces causing potato common scab in Western Europe. Plant Pathol 57(1):162–169

Frost KE, Groves RL, Charkowski AO (2013) Integrated control of potato pathogens through seed potato certification and provision of clean seed potatoes. Plant Dis 97(10):1268–1280. https://doi.org/10.1094/PDIS-05-13-0477-FE

Fry BA, Loria R (2002) Thaxtomin A: evidence for a plant cell wall target. Physiol Mol Plant Pathol 60(1):1–8

Fyans JK, Altowairish MS, Li Y, Bignell DRD (2015) Characterization of the coronatine-like phytotoxins produced by the common scab pathogen Streptomyces scabies. Mol Plant-Microbe Interact 28(4):443–454

Genin S, Denny TP (2012) Pathogenomics of the Ralstonia solanacearum species complex. Annu Rev Phytopathol 50:67–89

Goyer C, Beaulieu C (1997) Host range of Streptomycete strains causing common scab. Plant Dis 81(8):901–904

Goyer C, Faucher E, Beaulieu C (1996) Streptomyces Caviscabies sp. nov., from deep-pitted lesions in potatoes in Québec, Canada. Int J Syst Evol Microbiol 46(3):635–639

Guenthner J, Goolsby J, Greenway G (2012) Use and cost of insecticides to control potato psyllids and zebra chip on potatoes. Southwest Entomol 37(3):263–270

Han L, Dutilleul P, Prasher SO, Beaulieu C, Smith DL (2008) Assessment of common scab-inducing pathogen effects on potato underground organs via computed tomography scanning. Phytopathology 98(10):1118–1125

Hansen AK, Trumble JT, Stouthamer R, Paine TD (2008) A new Huanglongbing species,'Candidatus Liberibacter Psyllaurous,' found to infect tomato and potato, is vectored by the Psyllid Bactericera Cockerelli (Sulc). Appl Environ Microbiol 74(18):5862–5865

Hauben L, Moore ERB, Vauterin L, Steenackers M, Mergaert J, Verdonck L, Swings J (1998) Phylogenetic position of phytopathogens within the Enterobacteriaceae. Syst Appl Microbiol 21(3):384–397. https://doi.org/10.1016/S0723-2020(98)80048-9

Hayward AC (1964) Characteristics of Pseudomonas solanacearum. J Appl Bacteriol 27(2):265–277

Hellmers E (1958) Four wilt diseases of perpetual-flowering Carnations in Denmark. Dansk botanisk Arkiv 18(2):1–100. https://www.cabdirect.org/cabdirect/abstract/19581101855

Hill J, Lazarovits G (2005) A mail survey of growers to estimate potato common scab prevalence and economic loss in Canada. Can J Plant Pathol 27(1):46–52

Hiltunen LH, Alanen M, Laakso I, Kangas A, Virtanen E, Valkonen JPT (2011) Elimination of common scab sensitive progeny from a potato breeding population using Thaxtomin A as a selective agent. Plant Pathol 60(3):426–435

Hiltunen LH, Kelloniemi J, Valkonen JPT (2017) Repeated applications of a nonpathogenic Streptomyces strain enhance development of suppressiveness to potato common scab. Plant Dis 101(1):224–232

Holtsmark I, Takle GW, Brurberg MB (2008) Expression of putative virulence factors in the potato pathogen Clavibacter michiganensis subsp. sepedonicus during infection. Arch Microbiol 189(2):131–139

Huet G (2014) Breeding for resistances to Ralstonia Solanacearum. Front Plant Sci 5:715

Humphris SN, Cahill G, Elphinstone JG, Kelly R, Parkinson NM, Pritchard L, Toth IK, Saddler GS (2015) Detection of the bacterial potato pathogens Pectobacterium and Dickeya Spp. using conventional and real-time PCR. In: Lacomme C (ed) Plant pathology: techniques and protocols, Methods in molecular biology. Springer, New York, pp 1–16. https://doi.org/10.1007/978-1-4939-2620-6_1

Jacobs JM, Babujee L, Meng F, Milling A, Allen C (2012) The in planta transcriptome of Ralstonia solanacearum: conserved physiological and virulence strategies during bacterial wilt of tomato. MBio 3(4):e00114–e00112

Janse JD, Araluppan FAX, Schans J, Wenneker M, Westerhuis W (1998) Experiences with bacterial brown rot Ralstonia solanacearum biovar 2, race 3 in the Netherlands, Bacterial wilt disease. Springer, Berlin, pp 146–152

Joshi MV, Loria R (2007) Streptomyces turgidiscabies possesses a functional cytokinin biosynthetic pathway and produces leafy galls. Mol Plant-Microbe Interact 20(7):751–758

Khatri BB, Tegg RS, Brown PH, Wilson CR (2011) Temporal association of potato tuber development with susceptibility to common scab and Streptomyces scabiei-induced responses in the potato periderm. Plant Pathol 60(4):776–786

Khayi S, Cigna J, Chong TM, Quêtu-Laurent A, Chan K-G, Hélias V, Faure D (2016) Transfer of the potato plant isolates of Pectobacterium wasabiae to Pectobacterium parmentieri sp. nov. Int J Syst Evol Microbiol 66(12):5379–5383

Kim H-S, Ma B, Perna NT, Charkowski AO (2009) Phylogeny and virulence of naturally occurring type III secretion system-deficient Pectobacterium strains. Appl Environ Microbiol 75(13):4539–4549

King RR, Harold Lawrence C, Clark MC, Calhoun LA (1989) Isolation and characterization of phytotoxins associated with Streptomyces scabies. J Chem Soc Chem Commun 13:849–850

King RR, Harold Lawrence C, Clark MC (1991) Correlation of phytotoxin production with pathogenicity of streptomyces scabies isolates from scab infected potato tubers. Am Potato J 68(10):675–680

Kinkel LL, Bowers JH, Shimizu K, Neeno-Eckwall EC, Schottel JL (1998) Quantitative relationships among Thaxtomin A production, potato scab severity, and fatty acid composition in streptomyces. Can J Microbiol 44(8):768–776

Kloepper JW, Schroth MN (1981) Relationship of in vitro antibiosis of plant growth-promoting rhizobacteria to plant growth and the displacement of root microflora. Phytopathology 71: 1020–1024

Komeil D, Simao-Beaunoir A-M, Beaulieu C (2013) Detection of potential suberinase-encoding genes in Streptomyces scabiei strains and other actinobacteria. Can J Microbiol 59(5):294–303

Kong HG, Bae JY, Lee HJ, Joo HJ, Jung EJ, Chung E, Lee S-W (2014) Induction of the viable but nonculturable state of Ralstonia solanacearum by low temperature in the soil microcosm and its resuscitation by catalase. PLoS One 9(10):e109792

Kritzman G, Shani-Cahani A, Kirshner B, Riven Y, Bar Z, Katan J, Grinstein A (1996) Pod wart disease of peanuts. Phytoparasitica 24(4):293–304

Kumar GNM, Knowles LO, Richard Knowles N (2015) Zebra chip disease decreases tuber (Solanum tuberosum L.) protein content by attenuating protease inhibitor levels and increasing protease activities. Planta 242(5):1153–1166

Laine MJ, Haapalainen M, Wahlroos T, Kankare K, Nissinen R, Kassuwi S, Metzler MC (2000) The cellulase encoded by the native plasmid of Clavibacter michiganensis ssp. sepedonicus plays a role in virulence and contains an expansin-like domain. Physiol Mol Plant Pathol 57(5):221–233

Lambert DH, Loria R (1989a) Streptomyces scabies sp. nov., nom. rev. Int J Syst Evol Microbiol 39:387–392

Lambert DH, Loria R (1989b) Streptomyces acidiscabies sp. nov. Int J Syst Evol Microbiol 39:393–396

Larkin RP, Griffin TS (2007) Control of Soilborne potato diseases using Brassica green manures. Crop Prot 26(7):1067–1077

Larkin RP, Halloran JM (2014) Management effects of disease-suppressive rotation crops on potato yield and soilborne disease and their economic implications in potato production. Am J Potato Res 91(5):429–439

Larkin RP, Wayne Honeycutt C, Griffin TS, Modesto Olanya O, Halloran JM, He Z (2011) Effects of different potato cropping system approaches and water management on soilborne diseases and soil microbial communities. Phytopathology 101(1):58–67

Laurila J, Metzler MC, Ishimaru CA, Rokka V-M (2003) Infection of plant material derived from Solanum acaule with Clavibacter michiganensis ssp. sepedonicus: temperature as a determining factor in immunity of S. acaule to bacterial ring rot. Plant Pathol 52(4):496–504

Laurila J, Ahola V, Lehtinen A, Joutsjoki T, Hannukkala A, Rahkonen A, Pirhonen M (2008) Characterization of Dickeya strains isolated from potato and river water samples in Finland. Eur J Plant Pathol 122(2):213–225. https://doi.org/10.1007/s10658-008-9274-5

Lehtonen MJ, Rantala H, Kreuze JF, Bång H, Kuisma L, Koski P, Virtanen E, Vihlman K, Valkonen JPT (2004) Occurrence and survival of potato scab pathogens (Streptomyces species) on tuber lesions: quick diagnosis based on a PCR-based assay. Plant Pathol 53(3):280–287

Lévy JG, Scheuring DC, Koym JW, Henne DC, Tamborindeguy C, Pierson E, Creighton Miller J (2015) Investigations on putative zebra chip tolerant potato selections. Am J Potato Res 92(3):417–425

Lin H, Lou B, Glynn JM, Doddapaneni H, Civerolo EL, Chen C, Duan Y, Zhou L, Vahling CM (2011) The complete genome sequence of 'Candidatus Liberibacter solanacearum', the bacterium associated with potato zebra chip disease. PLoS One 6(4):e19135

Lu Y, Hatsugai N, Katagiri F, Ishimaru CA, Glazebrook J (2015) Putative serine protease effectors of Clavibacter michiganensis induce a hypersensitive response in the apoplast of Nicotiana species. Mol Plant-Microbe Interact 28(11):1216–1226

Ma B, Hibbing ME, Kim HS, Reedy RM, Yedidia I, Breuer J, Breuer J, Glasner JD, Perna NT, Kelman A, Charkowski AO (2007a) Host range and molecular phylogenies of the soft rot enterobacterial genera Pectobacterium and Dickeya. Phytopathology 97(9):1150–1163. https://apsjournals.apsnet.org/doi/abs/10.1094/PHYTO-97-9-1150

Ma B, Hibbing ME, Kim H-S, Reedy RM, Yedidia I, Breuer J, Breuer J, Glasner JD, Perna NT, Kelman A (2007b) Host range and molecular phylogenies of the soft rot enterobacterial genera Pectobacterium and Dickeya. Phytopathology 97(9):1150–1163

Mansfield J, Genin S, Magori S, Citovsky V, Sriariyanum M, Ronald P, Dow MAX, Verdier V, Beer SV, Machado MA (2012) Top 10 plant pathogenic bacteria in molecular plant pathology. Mol Plant Pathol 13(6):614–629

Manzer FE, McKenzie AR (1988) Cultivar response to bacterial ring rot infection in Maine. Am Potato J 65(6):333–339

Meng F (2013) The virulence factors of the bacterial wilt pathogen Ralstonia solanacearum. J Plant Pathol Microbiol 4(168):10–4172

Miyajima K, Tanaka F, Takeuchi T, Kuninaga S (1998) Streptomyces turgidiscabies sp. nov. Int J Syst Evol Microbiol 48(2):495–502

Nabhan S, De Boer SH, Maiss E, Wydra K (2013) Pectobacterium aroidearum sp. nov., a soft rot pathogen with preference for monocotyledonous plants. Int J Syst Evol Microbiol 63(7):2520–2525

Narusaka M, Kubo Y, Hatakeyama K, Imamura J, Ezura H, Nanasato Y, Tabei Y, Takano Y, Shirasu K, Narusaka Y (2013) Interfamily transfer of dual NB-LRR genes confers resistance to multiple pathogens. PLoS One 8(2):e55954

Natsume M, Komiya M, Koyanagi F, Tashiro N, Kawaide H, Abe H (2005) Phytotoxin produced by Streptomyces sp. causing potato russet scab in Japan. J Gen Plant Pathol 71(5):364–369

Natsume M, Tashiro N, Doi A, Nishi Y, Kawaide H (2017) Effects of concanamycins produced by Streptomyces scabies on lesion type of common scab of potato. J Gen Plant Pathol 83(2):78–82

Nelson WR, Fisher TW, Munyaneza JE (2011) Haplotypes of 'Candidatus Liberibacter solanacearum' suggest long-standing separation. Eur J Plant Pathol 130(1):5–12

Ngadze E, Brady CL, Coutinho TA, Van der Waals JE (2012) Pectinolytic bacteria associated with potato soft rot and blackleg in South Africa and Zimbabwe. Eur J Plant Pathol 134(3):533–549

Nissinen R, Kassuwi S, Peltola R, Metzler MC (2001) In planta-complementation of Clavibacter michiganensis subsp. sepedonicus strains deficient in cellulase production or HR induction restores virulence. Eur J Plant Pathol 107(2):175–182

Nykyri J, Niemi O, Koskinen P, Nokso-Koivisto J, Pasanen M, Broberg M, Plyusnin I, Törönen P, Holm L, Pirhonen M (2012) Revised phylogeny and novel horizontally acquired virulence determinants of the model soft rot phytopathogen Pectobacterium wasabiae SCC3193. PLoS Pathog 8(11):e1003013

Oniki M, Suzui T, Araki T, Sonoda RI, Chiba T, Takeda T (1986a) Causal agent of russet scab of potato. Bull Nat Inst of Agro Environ Sci 2:45–60

Oniki M, Suzui T, Araki T, Sonoda R, Chiba T, Takeda T (1986b) Causal agent of russet scab of potato [Streptomyces sp.]. Bull Nat Inst Agro-Environ Sci. http://agris.fao.org/agris-search/search.do?recordID=JP870464388

Padilla-Reynaud R, Simao-Beaunoir A-M, Lerat S, Bernards MA, Beaulieu C (2015) Suberin regulates the production of cellulolytic enzymes in Streptomyces scabiei, the causal agent of potato common scab. Microbes Environ 30(3):245–253

Park DH, Kim JS, Kwon SW, Wilson C, Yu YM, Hur JH, Lim CK (2003) Streptomyces luridiscabiei sp. nov., Streptomyces puniciscabiei sp. nov. and Streptomyces niveiscabiei sp. nov., which cause potato common scab disease in Korea. Int J Syst Evol Microbiol 53(6):2049–2054

Parkinson N, Stead D, Bew J, Heeney J, Tsror (Lahkim) L, Elphinstone J (2009) Dickeya species relatedness and clade structure determined by comparison of RecA sequences. Int J Syst Evol Microbiol 59(10):2388–2393. https://doi.org/10.1099/ijs.0.009258-0

Parkinson N, DeVos P, Pirhonen M, Elphinstone J (2014) Dickeya aquatica sp. nov., isolated from waterways. Int J Syst Evol Microbiol 64(7):2264–2266. https://doi.org/10.1099/ijs.0.058693-0

Pavlista AD (2005) Early-season applications of sulfur fertilizers increase potato yield and reduce tuber defects. Agron J 97(2):599–603

Pérombelon MCM (2002) Potato diseases caused by soft rot erwinias: an overview of pathogenesis. Plant Pathol 51(1):1–12

Pérombelon MCM, Kelman A (1987) Blackleg and other potato diseases caused by soft rot erwinias: proposal for revision of terminology. Plant Dis 71(3):283–285

Pirhonen M, Flego D, Heikinheimo R, Tapio Palva E (1993) A small diffusible signal molecule is responsible for the global control of virulence and exoenzyme production in the plant pathogen Erwinia carotovora. EMBO J 12(6):2467–2476

Pitman AR, Wright PJ, Galbraith MD, Harrow SA (2008) Biochemical and genetic diversity of pectolytic enterobacteria causing soft rot disease of potatoes in New Zealand. Australas Plant Pathol 37(6):559–568

Pitman AR, Harrow SA, Visnovsky SB (2010) Genetic characterisation of Pectobacterium wasabiae causing soft rot disease of potato in New Zealand. Eur J Plant Pathol 126(3):423–435

Powelson ML, Rowe RC (2008) Managing diseases caused by seedborne and soilborne fungi and fungus-like pathogens. Potato Health Manage:2183–2195

Prior P, Ailloud F, Dalsing BL, Remenant B, Sanchez B, Allen C (2016) Genomic and proteomic evidence supporting the division of the plant pathogen Ralstonia solanacearum into three species. BMC Genomics 17(1):90

Qian Y-l, Wang X-s, Wang D-z, Zhang L-n, Chao-long Z, Gao Z-l, Zhang H-j, Wang Z-y, Sun X-y, Yao D-n (2013) The detection of QTLs controlling bacterial wilt resistance in tobacco (N. Tabacum L.). Euphytica 192(2):259–266

Qu X, Wanner LA, Christ BJ (2008) Using the TxtAB operon to quantify pathogenic Streptomyces in potato tubers and soil. Phytopathology 98(4):405–412

Rashed A, Workneh F, Paetzold L, Rush CM (2015) Emergence of 'Candidatus Liberibacter solanacearum'-infected seed potato in relation to the time of infection. Plant Dis 99(2):274–280

Rashidi M, Novy RG, Wallis CM, Rashed A (2017) Characterization of host plant resistance to zebra chip disease from species-derived potato genotypes and the identification of new sources of zebra chip resistance. PLoS One 12(8):e0183283

Rietman H, Finkers R, Evers L, van der Zouwen PS, van der Wolf JM, Visser RGF (2014) A stringent and broad screen of Solanum spp. tolerance against Erwinia bacteria using a petiole test. Am J Potato Res 91(2):204–214

Rossmann S, Dees MW, Perminow J, Meadow R, Brurberg MB (2018) Soft rot Enterobacteriaceae are carried by a large range of insect species in potato fields. Appl Environ Microbiol 84(12):e00281–e00218. https://doi.org/10.1128/AEM.00281-18

Safni I, Cleenwerck I, De Vos P, Fegan M, Sly L, Kappler U (2014) Polyphasic taxonomic revision of the Ralstonia solanacearum species complex: proposal to emend the descriptions of *Ralstonia solanacearum* and *Ralstonia syzygii* and reclassify current *R. syzygii* strains as Ralstonia syzygii subsp. syzygii subsp. nov., R. solanacearum phylotype IV strains as *Ralstonia syzygii* subsp. Indonesiensis subsp. nov., Banana blood disease bacterium strains as *Ralstonia syzygii* subsp. Celebesensis subsp. nov. and *R. solanacearum* phylotype I and III strains as *Ralstonia pseudosolanacearum* sp. nov. Int J Syst Evol Microbiol 64(9):3087–3103

Salgon S, Jourda C, Sauvage C, Daunay M-C, Reynaud B, Wicker E, Dintinger J (2017) Eggplant resistance to the Ralstonia solanacearum species complex involves both broad-spectrum and strain-specific quantitative trait loci. Front Plant Sci 8:828

Samson R, Legendre JB, Christen R, Saux MF-L, Achouak W, Gardan L (2005) Transfer of Pectobacterium chrysanthemi (Burkholder et Al. 1953) Brenner et Al. 1973 and Brenneria paradisiaca to the genus Dickeya gen. nov. as Dickeya chrysanthemi comb. nov. and Dickeya paradisiaca comb. nov. and delineation of four novel species, Dickeya dadantii sp. nov., Dickeya dianthicola sp. nov., Dickeya dieffenbachiae sp. nov. and Dickeya zeae sp. nov. Int J Syst Evol Microbiol 55(Pt 4):1415–1427. https://doi.org/10.1099/ijs.0.02791-0.

Santos-Cervantes ME, Felix-Gastelum R, Herrera-Rodríguez G, Espinoza-Mancillas MG, Mora-Romero AG, Leyva-López NE (2017) Characterization, pathogenicity and chemical control of Streptomyces acidiscabies associated to potato common scab. Am J Potato Res 94(1):14–25

Sarfraz S, Riaz K, Oulghazi S, Cigna J, Sahi ST, Khan SH, Faure D (2018) Pectobacterium punjabense sp. nov., isolated from blackleg symptoms of potato plants in Pakistan. Int J Syst Evol Microbiol 68(11):3551–3556

Sarikhani E, Sagova-Mareckova M, Omelka M, Kopecky J (2017) The effect of peat and iron supplements on the severity of potato common scab and bacterial community in tuberosphere soil. FEMS Microbiol Ecol 93(1)

Schlatter D, Kinkel L, Thomashow L, Weller D, Paulitz T (2017) Disease suppressive soils: new insights from the soil microbiome. Phytopathology 107(11):1284–1297

Secor GA, Rivera VV, Abad JA, Lee I-M, Clover GRG, Liefting LW, Li X, De Boer SH (2009) Association of 'Candidatus Liberibacter solanacearum'with zebra chip disease of potato established by graft and psyllid transmission, electron microscopy, and PCR. Plant Dis 93(6):574–583

Sengoda VG, Rodney Cooper W, Swisher KD, Henne DC, Munyaneza JE (2014) Latent period and transmission of 'Candidatus Liberibacter solanacearum' by the potato Psyllid Bactericera cockerelli (Hemiptera: Triozidae). PLoS One 9(3):e93475

She XM, Lan GB, Tang YF, He ZF (2017) Pectrobacterium caratovorum subsp. Brasilicnsc causing pepper black spot disease in China. J Plant Pathol 99(3):769–772

Sławiak M, van Beckhoven JRCM, Speksnijder AGCL, Czajkowski R, Grabe G, van der Wolf JM (2009) Biochemical and genetical analysis reveal a new clade of biovar 3 Dickeya spp. strains isolated from potato in Europe. Eur J Plant Pathol 125(2):245–261. https://doi.org/10.1007/s10658-009-9479-2

Teresani GR, Bertolini E, Alfaro-Fernández A, Martínez C, Tanaka FAO, Kitajima EW, Roselló M, Sanjuán S, Ferrándiz JC, López MM (2014) Association of 'Candidatus Liberibacter solanacearum'with a vegetative disorder of celery in Spain and development of a real-time PCR method for its detection. Phytopathology 104(8):804–811

Teulon DAJ, Workman PJ, Nielsen MC, Thomas KL, Anderson DP (2009) Bactericera cockerelli: incursion, dispersal and current distribution on vegetable crops in New Zealand. New Zealand Plant Prot 62:136–144

Thangavel T, Tegg RS, Wilson CR (2016) Toughing it out—disease-resistant potato mutants have enhanced tuber skin defenses. Phytopathology 106(5):474–483

Thaxter R (1892) Potato scab. Potato Scab:153–160. https://www.cabdirect.org/cabdirect/abstract/20057000599

Thinakaran J, Pierson E, Kunta M, Munyaneza JE, Rush CM, Henne DC (2015) Silverleaf nightshade (Solanum Elaeagnifolium), a reservoir host for 'Candidatus Liberibacter solanacearum', the putative causal agent of zebra chip disease of potato. Plant Dis 99(7):910–915

Tian Y, Zhao Y, Yuan X, Yi J, Fan J, Zhigang X, Baishi H, De Boer SH, Li X (2016) Dickeyafangzhongdai sp. nov., a plant-pathogenic bacterium isolated from pear trees (Pyrus Pyrifolia). Int J Syst Evol Microbiol 66(8):2831–2835. https://doi.org/10.1099/ijsem.0.001060

Tomihama T, Nishi Y, Mori K, Shirao T, Iida T, Uzuhashi S, Ohkuma M, Ikeda S (2016) Rice bran amendment suppresses potato common scab by increasing antagonistic bacterial community levels in the rhizosphere. Phytopathology 106(7):719–728

Toth IK, van der Wolf JM, Saddler G, Lojkowska E, Hélias V, Pirhonen M, Tsror (Lahkim) L, Elphinstone JG (2011) Dickeya species: an emerging problem for potato production in Europe. Plant Pathol 60(3):385–399. https://doi.org/10.1111/j.1365-3059.2011.02427.x

van der Merwe JJ, Coutinho TA, Korsten L, van der Waals JE (2010) Pectobacterium carotovorum subsp. brasiliensis causing blackleg on potatoes in South Africa. Eur J Plant Pathol 126(2):175–185

Van der Wolf JM, Elphinstone JG, Stead DE, Metzler M, Müller P, Hukkanen A, Karjalainen R (2005a) Epidemiology of Clavibacter michiganensis subsp. sepedonicus in relation to control of bacterial ring rot. PRI Bioscience (Report/Plant Research International 95)—38

Van der Wolf JM, Van Beckhoven JRCM, Hukkanen A, Karjalainen R, Müller P (2005b) Fate of *Clavibacter michiganensis* ssp. sepedonicus, the causal organism of bacterial ring rot of potato, in weeds and field crops. J Phytopathol 153(6):358–365

Van der Wolf JM, de Haan EG, Kastelein P, Krijger M, de Haas BH, Velvis H, Mendes O, Kooman-Gersmann M, van der Zouwen PS (2017) Virulence of Pectobacterium carotovorum subsp. brasiliense on potato compared with that of other Pectobacterium and Dickeya species under climatic conditions prevailing in the Netherlands. Plant Pathol 66(4):571–583

Van Vaerenbergh J, De Paepe B, Hoedekie A, Van Malderghem C, Zaluga J, De Vos P, Maes M (2016) Natural infection of *Clavibacter michiganensis* subsp. sepedonicus in tomato (Solanum Lycopersicum). New Dis Rep 33:7–7

Waldee EL (1942) Comparative studies of some peritrichous phytopathogenic bacteria. Retrospective Theses and Dissertations. p 14150. https://lib.dr.iastate.edu/rtd/14150

Waleron M, Misztak A, Waleron M, Franczuk M, Wielgomas B, Waleron K (2018) Transfer of Pectobacterium carotovorum subsp. carotovorum strains isolated from potatoes grown at high altitudes to *Pectobacterium peruviense* sp. nov. Syst Appl Microbiol 41(2):85–93

Wallis CM, Munyaneza JE, Chen J, Novy R, Bester G, Buchman JL, Nordgaard J, Hest PV (2015) 'Candidatus Liberibacter solanacearum' titers in and infection fffects on potato tuber chemistry of promising germplasm exhibiting tolerance to zebra chip disease. Phytopathology 105(12):1573–1584

Wallis CM, Rashed A, Workneh F, Paetzold L, Rush CM (2017) Effects of holding temperatures on the development of zebra chip symptoms,'Candidatus Liberibacter solanacearum' titers, and phenolic levels in 'Red La Soda' and 'Russet Norkotah' tubers. Am J Potato Res 94(4):334–341

Wang J-F, Ho F-I, Truong HTH, Huang S-M, Balatero CH, Dittapongpitch V, Hidayati N (2013) Identification of major QTLs associated with stable resistance of tomato cultivar 'Hawaii 7996'to Ralstonia solanacearum. Euphytica 190(2):241–252

Wang J, Haapalainen M, Schott T, Thompson SM, Smith GR, Nissinen AI, Pirhonen M (2017a) Genomic sequence of 'Candidatus Liberibacter solanacearum' haplotype C and its comparison with haplotype A and B genomes. PLoS One 12(2):e0171531

Wang N, Pierson EA, Setubal JC, Xu J, Levy JG, Zhang Y, Li J, Rangel LT, Jr JM (2017b) The Candidatus Liberibacter–host interface: insights into pathogenesis mechanisms and disease control. Annu Rev Phytopathol 55:451–482

Wang J, Wang YH, Dai PG, Chen DX, Zhao TC, Li XL, Huang Q (2017c) First report of tobacco bacterial leaf blight caused by Pectobacterium carotovorum subsp. Brasiliense in China. Plant Dis 101(5):830–830

Wanner LA (2006) A survey of genetic variation in Streptomyces isolates causing potato common scab in the United States. Phytopathology 96(12):1363–1371

Wanner LA (2007) High proportions of nonpathogenic Streptomyces are associated with common scab-resistant potato lines and less severe disease. Can J Microbiol 53(9):1062–1075

Wanner LA (2009) A patchwork of Streptomyces species isolated from potato common scab lesions in North America. Am J Potato Res 86(4):247–264

Williamson L, Nakaho K, Hudelson B, Allen C (2002) Ralstonia solanacearum race 3, biovar 2 strains isolated from geranium are pathogenic on potato. Plant Dis 86(9):987–991

Winslow C-EA, Jean B, Buchanan RE, Krumwiede C Jr, Rogers LA, Smith GH (1917) The families and genera of the bacteria: preliminary report of the Committee of the Society of American Bacteriologists on characterization and classification of bacterial types. J Bacteriol 2(5):505

Yang FY, Yang DJ, Zhao WQ, Liu DQ, Yu XM (2017) First report of Streptomyces diastatochromogenes causing potato common scab in China. Plant Dis 101(1):243–243

Yao J, Saenkham P, Levy J, Ibanez F, Noroy C, Mendoza A, Huot O, Meyer DF, Tamborindeguy C (2016) Interactions 'Candidatus Liberibacter solanacearum'—Bactericera cockerelli: haplotype effect on vector fitness and gene expression analyses. Front Cell Infect Microbiol 6:62

Yap M-N, Barak JD, Charkowski AO (2004) Genomic diversity of Erwinia carotovora subsp. carotovora and its correlation with virulence. Appl Environ Microbiol 70(5):3013–3023

Zhang Y, Loria R (2017) Emergence of novel pathogenic Streptomyces species by site-specific accretion and Cis-mobilization of Pathogenicity Islands. Mol Plant-Microbe Interact 30(1):72–82

Zhang Y, Bignell DRD, Zuo R, Fan Q, Huguet-Tapia JC, Ding Y, Loria R (2016a) Promiscuous pathogenicity islands and phylogeny of pathogenic Streptomyces spp. Mol Plant-Microbe Interact 29(8):640–650

Zhang Y, Fan Q, Loria R (2016b) A re-evaluation of the taxonomy of phytopathogenic genera Dickeya and Pectobacterium using whole-genome sequencing data. Syst Appl Microbiol 39(4):252–259

5

Insect Pests Affecting Potatoes in Tropical, Subtropical, and Temperate Regions

Jürgen Kroschel, Norma Mujica, Joshua Okonya, and Andrei Alyokhin

Abstract Ensuring the sustainable production of potato is an important challenge facing agriculture globally. Insect pests are major biotic constraints affecting potato yields and tuber quality. The high pesticide uses to control them is of high human and environmental health concern, and it is expected that this will be further exacerbated through impacts of climate change. The chapter provides an overview of the geographical distribution of potato insect pests and their importance in tropical, subtropical, and temperate potato production regions. Climate change will potentially contribute to expand their geographical range of distribution, and increasing populations will lead to greater crop and post-harvest losses. Good progress has been made in applying insect pest modeling in pest risk analysis of potato pests to inform and create better awareness of future pest risks under climate change. Potato pests include some of the species which have evolved resistance to a wide variety of chemicals; and potato growers have already experienced the situation that available chemicals failed to control their targets. This chapter emphasizes the development, use, and adaptation of Integrated Pest Management (IPM) across all potato-growing regions of the world. Ultimately, this will lead to sustainable and more resilient potato production systems not overly dependent on pesticides. IPM requires a good knowledge and understanding of individual potato production systems; identifying pest species, knowing their biology and symptoms of infestation is essential for making educated decisions on their integrated management. To address this need,

J. Kroschel (✉)
Independent Consultant for International Agricultural Research & Development, Filderstadt, Germany
e-mail: jurgen.kroschelconsult@outlook.com

N. Mujica
Independent Consultant, Lima, Peru

J. Okonya
International Potato Center, Kampala, Uganda
e-mail: J.Okonya@cgiar.org

A. Alyokhin
University of Maine, Orono, ME, USA
e-mail: alyokhin@maine.edu

the chapter provides detailed information for a total of 49 insect pests of potato and the status quo of their management around the world.

Introduction

The potato (*Solanum tuberosum* L.) is native to the Andean highlands of South America. Today, potato is produced in more than 149 countries in temperate, subtropical, and tropical agroecologies, demonstrating the versatility and adaptability of this crop to a wide range of environmental conditions. Potato provides humans with an abundant and relatively inexpensive source of high-quality nutrients. Historically, their incorporation into everyday diets commonly coincided with periods of rapid population growth in a variety of nations (Zuckerman 1999). Presently, potatoes continue playing a very important role in feeding the human population. The last 30 years were characterized by an explosive growth in their popularity in Asia and Africa, which were previously reliant on other staple crops. Therefore, ensuring sustainability of potato production is currently an important challenge facing agricultural professionals worldwide (Vincent et al. 2013). Insect pests are major biotic factors affecting potato yield and tuber quality. Globally, losses are estimated on average at 16% (Oerke et al. 1994). Locally, if not routinely controlled, reductions in tuber yield and quality can be between 30 and 70% for various pests (Mujica and Kroschel 2013; Kroschel and Schaub 2013). The high pesticide use in potato is of high human and environmental health concern, which needs to be addressed by developing and more widely implementing Integrated Pest Management (IPM) approaches.

Potato Insect Pests' Geographical Distribution and Invasiveness

Due its global geographical distribution, potato is affected by a wide range of insect pests. In this book chapter we listed and described a total of 49 species: nine major species occurring in tropical and subtropical regions; two major species affecting potato in temperate regions; six major and 32 minor species of temperate, subtropical, and tropical regions. Farmers in tropical and subtropical countries must contend with a higher number of pest species, and with some exceptions, a minimum of 2–4 pests often reach pest status requiring the application of control methods (Kroschel et al. 2012). Many pests have evolved in the center of potato origin, and farmers in the Andean region are confronted by a higher number of pests than farmers in Africa or Asia. Some species such as the potato tuber moth, *Phthorimaea operculella* (Zeller), and the leafminer fly, *Liriomyza huidobrensis* (Blanchard) have become invasive and occur today as serious pests in many tropical and subtropical regions.

In contrast, the strong adaptation of Andean potato weevils, *Premnotrypes* spp., to the climate of the Andean region and its monophagous feeding habitat on potato and its wild relatives have restricted its distribution. There are, however, still several other pests which could also gain global proportions. Just very recently in 2006, the tomato leaf miner, *Tuta absoluta* Meyrick, although a more minor pest in potato, was unintentionally introduced to Spain, from where it continued its devastating journey across Africa and into Asia where it reached India within less than 10 years. As farmers had not been prepared and no control measures had been in place, the pest caused large production losses in tomato (*Lycopersicon esculentum* Mill.); under certain conditions also potato was more heavily infested as known from South America. The bud midge, *Prodiplosis longifilia* Gagne, currently with a restricted distribution in Florida and Virginia, and South America (Colombia, Peru, and Ecuador) could become an invasive species supported by its very polyphagous feeding habit. The Colorado potato beetle, *Leptinotarsa decemlineata* (Say), native to Mexico, has spread across most of the United States, and was introduced into France in the 1920s from where it spread further reaching also parts of China (CABI 2017a).

Impacts of Climate Change on Potato Insect Pests

Climate, especially temperature, has a strong and direct influence on the development and growth of insect pest populations. Herbivorous insects—as all other arthropods—are exothermic organisms that cannot internally regulate their own temperature. Their development depends on the temperature to which they are exposed in the environment. A rise in temperature due to climate change may both increase or decrease pest development rates and related crop losses. Hence, an increase in temperature can potentially affect range expansion and outbreaks of many insect pests including pests of potato.

Innovative modeling approaches such as process-based climatic response phenology models (Orlandini et al. 2017; Mujica et al. 2017; Sporleder et al. 2004, 2017) have been used to assess the effect of temperature increase under projected changes in global temperature for the year 2050 (Kroschel et al. 2016a) for a wide range of potato pests: *P. operculella* (Kroschel et al. 2013, 2016b); *L. huidobrensis* (Mujica et al. 2016); Guatemalan potato tuber moth, *Tecia solanivora* (Povolny) (Schaub et al. 2016); Andean potato tuber moth, *Symmetrischema tangolias* (Gyen) (Sporleder et al. 2016); the White flies, *Bemisia tabaci* (Gennadius) and *Trialeurodes vaporariorum* (Westwood) (Gamarra et al. 2016a, b).

These predictions have clearly demonstrated that insect pests of potato will respond to climate change by expanding their geographical range of distribution and increasing population densities will lead to greater crop and post-harvest losses. There are, however, distinct differences among the different pest species. The damage potential of *P. operculella* for example will potentially progressively increase in all regions where the pest already prevails today, with a range expansion into temperate and tropical mountainous regions (Kroschel et al. 2013, 2016b). In comparison, *L. huidobrensis* is much less adapted to warmer climates and climate change will differently affect this species. The global predictions clearly indicate a slight to moderate decrease in the establishment potential of the pest in most of the tropical and subtropical potato production areas, but still with high pest risks. Further, a range expansion into temperate regions of Asia, North and South America, and Europe, as well as into subtropical and tropical mountainous regions is expected, with a moderate increase of its establishment and damage potential (Mujica et al. 2016). A further range expansion into tropical mountainous regions such as the Andes has been also predicted for *T. solanivora* and *S. tangolias* (Quiroz et al. 2018).

Potato production systems of tropical countries are highly susceptible to pest infestations due to often year-round favorable climatic conditions for pest population growth and host plant availability. Even smaller changes in temperature predicted for tropical regions compared to temperate regions will have stronger consequences on pest development due to already higher existing metabolism rates of organisms such as insects (Dillon et al. 2010). This does affect not only the general life cycle of an insect pest but also all other biological processes, including feeding rates, plant growth, and activity of biotic antagonist. Lessons have been learnt from the El Niño phenomena to better understand possible climate change effects on pest abundance and severity in tropical production areas. During the 1997 El Niño phenomena in Peru, the mean temperature on the Peruvian coast increased by about 5 °C above the annual average. While infestation of potato by *L. huidobrensis* decreased, infestations by all other pests increased in agricultural and horticultural crops. The farmers' only adaptive strategy to cope was to apply high doses of pesticides every 2–3 days (Cisneros and Mujica 1999a).

Insect Pest Control with Insecticides in Potato

Potato foliage contains a considerable amount of glycoalkaloids, which provide at least some protection from herbivory. Nevertheless, they are attacked by a robust complex of phytophagous insects, some of which can destroy the potato crop in the absence of adequate control measures. Just as with most other cultivated plants, management of insect pests of potato is achieved predominantly through application of pesticides. By some estimates, potatoes are the most chemically dependent crop in the world (Vincent et al. 2013). Insect pests are not the only factor responsible for this notoriety, as large amounts of fungicides are used to combat diseases caused by fungi and oomycetes. Still, insecticides remain to be a foundation of insect pest management in most potato fields around the world, and their use can be rather heavy at times.

Although insecticides have been largely successful in keeping potato production going, there are serious and well-known concerns about long-term sustainability of this approach. Nontarget effects of insecticides on a variety of organisms, including humans and beneficial insects, gained considerable notoriety since 1960s. The use of highly hazardous pesticides in potato in countries such as Ecuador and Peru has caused serious health risks to farmers (Orozco et al. 2009). Worldwide decline in beneficial pollinators documented in the early 2000s provided additional fuel to the fire of public apprehension of using toxic chemicals in agriculture. Furthermore, as discussed in the following section, more and more insecticides have lost their efficiency due to resistance development in insect populations (Alyokhin et al. 2013, 2015).

As more and more insecticides are becoming phased out due to environmental concerns or become ineffective due to resistance development in targeted insect populations, the number of options available to potato growers dwindles. Developing replacement insecticides is an increasingly difficult and expensive task, and it is highly questionable that a plethora of new active ingredients will regularly appear on the market in perpetuity (Alyokhin et al. 2015). Therefore, good stewardship of existing chemicals and, whenever possible, their replacement with nonchemical control alternatives become an increasingly important business strategy for the pesticide industry and potato farmers. Therefore, this chapter puts a considerable emphasis on describing various nonchemical management options for insect pests.

In the modern age of industrialized agriculture, a farmer field is often considered to be a type of a production facility. Although there is a certain element of truth to such an approach, it is important to remember that it is still comprised of living organisms that are interacting with each other and with their environment. In other words, it is an ecosystem, with the same basic characteristics as all other ecosystems on Earth. This includes evolutionary processes leading to adaptation to a particular set of environmental conditions through survival and reproduction of specific genotypes. In many cases, such an adaptation means developing an ability to survive exposure to insecticides, a phenomenon known as insecticide resistance (Alyokhin et al. 2015).

Potato pests include some of the species that are most prone to evolving resistance to a wide variety of chemicals. The Arthropod Pesticide Resistance Database (2018) lists 469 cases of green peach aphid (*Myzus persicae* (Sulzer)) resistance to a total of 80 active ingredients; 300 cases of Colorado potato beetle (*Leptinotarsa decemlineata* (Say)) resistance to a total of 56 active ingredients; 111 cases of greenhouse whitefly (*T. vaporariorum*) resistance to 27 active ingredients; and 501 cases of two-spotted spider mite (*Tetranychus urticae* C. L. Koch) resistance to rather impressive 95 active ingredients. The extent of resistance is likely to be underestimated because not every case of its development is entered into the database. It is possible that the ability to deal with toxic glycoalkoloids contained in potato foliage serves as a preadaptation to resisting chemical toxins made by humans (Alyokhin and Chen 2017). Not every population of a given pest species is resistant to all compounds that have been recorded to fail against that species. However, these statistics vividly illustrate the seriousness of the problem. On several occasions, potato growers already experienced the situation when virtually all commercially available chemicals failed to control their targets (Alyokhin et al. 2013). It is very likely that such a situation will arise again in the foreseeable future. Therefore, we need to be proactive to prevent this from happening.

Insecticide resistance is usually associated with several traits that may be taken advantage of when devising resistance management plans. Due to the pleiotropic effects of resistant alleles, in the absence of insecticides resistant insects often have lower reproductive output and/or suffer higher mortality. As a result, they are being outcompeted by susceptible insects. Also, in many cases resistance is inherited as an incompletely dominant or an incompletely recessive trait. Therefore, insects that are heterozygous at the resistant locus can be successfully controlled with sufficiently high dose of insecticide. So, resistance management techniques are generally directed towards preventing the situation when only highly resistant homozygotes survive in a population (Alyokhin et al. 2013, 2015). In practical terms, this can be achieved by doing the following:

- Monitoring insecticide efficacy. Larger than usual number of surviving pests may indicate that their population is becoming resistant.
- Avoiding applications of the same or related products repeatedly throughout a growing season. Instead, application schedule should consist of a sequence of insecticides with different modes of action. Information on a product's mode of action is available from its manufacturer.
- Applying insecticides at rates that are not lower than a recommended minimum. Otherwise, heterozygotes will survive and breed with each other. Following Mendelian laws of inheritance, some of their offspring will become homozygously resistant and capable of surviving even the recommended insecticide rates.
- Applying insecticides only when pest populations are sufficiently high to cause economically important damage. Therefore, the use of control thresholds developed for different potato pests is highly recommended. Trying to kill every single insect will, by definition, result in the survival of only highly resistant genotypes.
- Whenever possible, leaving parts of the field untreated to allow susceptible pests to survive and interbreed with resistant pests. Because resistance is almost never completely dominant, resulting heterozygotes will be killed by insecticides.
- The use of insecticide applications should be not the first but the ultimate control option in an IPM approach after all other management options could not prevent to keep a specific pest population under the economic threshold.

Integrated Pest Management in Potato

Integrated Pest Management (IPM) is defined as an "ecosystem approach to crop production and protection that combines different management strategies and practices to grow healthy crops and minimize the use of pesticides." It means "a careful consideration of all available pest control techniques and subsequent integration of appropriate measures that discourage the development of pest populations and keep pesticides and other interventions to levels that are economically justified and reduce or minimize risks to human health and the environment. IPM emphasizes the growth of a healthy crop with the least possible disruption to agro-ecosystems and encourages natural pest control mechanisms" (FAO 2018). It has been also defined as "a decision support system for the selection and use of pest control tactics, singly or harmoniously coordinated into a management strategy, based on cost/benefit analyses that consider the interests of and impacts on producers, society, and the environment" (Kogan 1998).

Although IPM has been widely promoted by researchers, policymakers, and agricultural practitioners, it is still very far from being universally adopted, but there have been large shifts towards bringing it into the mainstream of agricultural production. Since 2014, for example, a European Union (EU) Directive has obliged all professional plant growers within the Union to apply the general principles of IPM (European Parliament and the Council of the European Union 2009). To make IPM, however, successful, integrating several methods to combat pests requires interdisciplinary research and effective nonchemical control methods for all pests in a specific crop agroecology to fully avoid or minimize the pesticide use. But many research programs working on IPM still focus on single plant protection methods and not on a systematic study of the compatibility and optimization of simultaneously implemented pest management elements (Stenberg 2017). The International Potato Center (CIP) has developed a holistic working framework for potato pest management research and development and its application by farmers. This approach led, for example, to the development of IPM for different potato agroecologies in Peru under smallholder production considering ecological, economic, and environmental benefits (Mujica and Kroschel 2018; Kroschel et al. 2012). Further, Horn and Page (2008) reported from a successful potato IPM program for large-scale potato growers in Australia.

Individual potato IPM programs vary greatly depending on the specific agroecology and the socioeconomic conditions for which it is developed or adapted. However, they all share several important components, which distinguish them from the so-called conventional approaches that largely rely on calendar-based sprayings of pesticides.

- Application of best cultural practices, which includes the use of healthy seed, suitable crop rotations, and intercropping systems among others which also support natural biological control.
- Correct and timely pest identification and regular monitoring of population development. This applies the use of monitoring tools such as pheromone- or physical-based trapping methods (e.g., yellow sticky traps). Educated pest management takes advantage of specific aspects of pest biology that can be used to diminish its populations in one way or another. It should also be preventative, initiated before the occurrence of economic losses.
- Application of economic thresholds. Attempts on complete pest elimination are costly, often futile, and generally counterproductive. A control action should be

taken only when the cost of damage due to injury by a pest exceeds the cost of the control action. Control thresholds have been developed for several potato pests, but they need to be verified and adjusted for the conditions of a specific location and potato variety.

- Using diverse control tactics in a complementary or, whenever possible, synergistic ways. Pesticides are the ultimate option integrated with cultural, biological, regulatory, and other controls to result in a unified multi-pronged attack against pest populations.
- Evaluating results. This component is often overlooked, but very important. Efficiency of IPM control programs should be continuously monitored to make necessary adjustments when needed considering economic, ecological, and environmental assessments.

Among other benefits, IPM is likely to dramatically slow down evolution of insecticide resistance. Simultaneous adaptations to diverse and unrelated management techniques will require statistically unlikely genetic changes in pest populations (Alyokhin et al. 2015). As a result, integrating different techniques is likely to reduce our reliance on the "pesticide treadmill" of constantly replacing failed chemicals. With all its advantages, IPM should be made available to farmers across all potato-growing areas of the world. Practicing this approach requires, however, a good understanding of individual production systems down to a single-field level and knowing their specific components. Identifying pest species and knowing their biology is essential for making educated decisions on their management. To address this need, the following subchapters provide an overview of major and minor insect pests of potato and their management around the world.

Major Pests in Tropical and Subtropical Regions

Potato Tuber Moths

Phthorimaea operculella (Zeller, 1873), *Symmetrischema tangolias* (Gyen, 1913), *Tecia solanivora* (Povolny, 1973) (Lepidoptera: Gelechiidae)

Distribution The potato tuber moth (PTM), *Phthorimaea operculella,* originated in the tropical mountainous regions of South America. Today it has a worldwide distribution and is considered the most damaging potato pest in the developing world. It is present in almost all tropical and subtropical regions of the world, in North, Central, and South America, Africa, Asia, Australia, and Europe. The Andean potato tuber moth (APTM), *Symmetrischema tangolias,* is native to South America (Peru and Bolivia) but has in the last decades spread to other regions of the world. Records include North America, Australia, New Zealand, and more recently Indonesia. Although this species is known as a pest of potato and tomato in South America and the Australian region, it does not attack any of these

Fig. Adult female of *Phthorimaea operculella* (**a**) and symptoms of larvae infestation on leaves, stems, and tubers (**b**, **c**, **d**). (Photo credits: CIP)

Fig. Adult male and female of *Symmetrischema tangolias* (**a**), and symptoms of larvae infestation on stems and tubers (**b**, **c**). (Photo credits: CIP)

two crops in North America but instead feeds on black nightshade (*Solanum americanum* Mill.). The Guatemalan potato tuber moth (GPTM), *Tecia solanivora,* probably originated from Guatemala and is endemic throughout Central America . In 1983, the pest was unintentionally introduced into Venezuela and then invaded Colombia and Ecuador. In 2000, *T. solanivora* was introduced in the Canary Islands (Tenerife). Since then the pest has been considered as a major threat to potato throughout southern Europe and was listed as a quarantine pest by the European and Mediterranean Plant Protection Organization (EPPO 2005a; Kroschel et al. 2016a, b; Kroschel and Schaub 2013). In 2014, it was finally recorded in mainland Spain where efforts are going on to eradicate the pest (Jeger et al. 2018).

Host range *P. operculella* is an oligophagous pest (i.e., an insect feeding on a restricted range of food plants) of vegetable crops that belong mainly to the family Solanaceae: potato (*Solanum tuberosum* L.), tomato (*Lycopersicon esculentum* Mill.), and tobacco (*Nicotana tabacum* L.). Also, wild species of the Solanaceae family, including important weeds, e.g., black night shade (*Solanum nigrum* L.) are hosts. In total, the host range comprises 60 species. Further, crops of the family Chenopodiaceae are attacked including eggplant (*Solanum melongena* L.), bell

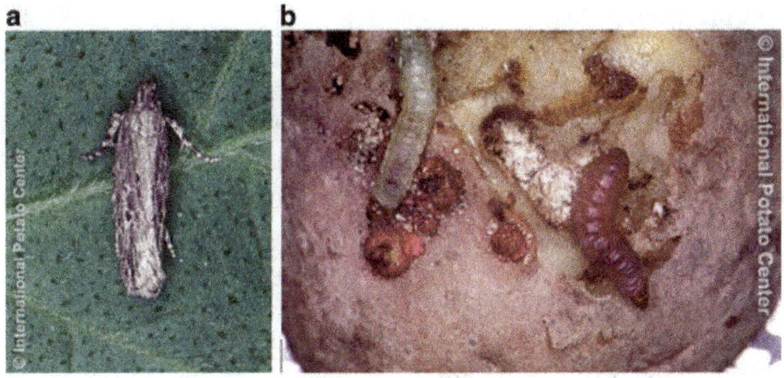

Fig. Adult male of *Tecia solanivora* (**a**), and symptoms of larvae infestation on tubers (**b**). (Photo credits: CIP)

pepper (*Capsium annuum* L.), Cape gooseberry (*Physalis peruviana* L.), aubergine (S. *melongena* L.), and sugar beet (*Beta vulgaris* L.). *S. tangolias* has a more restricted host range comprising of potato, tomato, sweet cucumber (*Solanum muricatum* Aiton), poroporo (*Solanum aviculare* G. Forst), kangaroo apple (*Solanum laciniatum* Aiton), black nightshade, and bell pepper. *T. solanivora* is monopahagous attacking only potato.

Symptoms of infestation *P. operculella* attacks potato by mining the leaves and stems and by feeding on the tuber. Mines are the typical symptoms of leaf damage caused by the larvae eating the mesophyll without damaging the upper and lower epidermis. When the foliage dies, the larvae enter the soil through cracks where they may eventually find and feed upon tubers. Larvae enter potato tubers via the eyes and continue to bore or tunnel through the tuber just below the skin. Larval excreta are pushed out through the holes, which can be observed immediately after larvae start their mining activity. Larvae of APTM enter the potato stem making a small hole in the plant axils (between stem and lateral petioles). From this hole, galleries made by the larvae run downward within the stem. Excreta are pushed out through the initial hole made by the larva. When stems are severely damaged, the upper part of the stem wilts or the whole plant collapses. Young plants can suffer tip death from boring larvae. Eggs may be found in slits on the stem of a food plant. In tubers, larvae enter through potato eyes. Initially, the small hole can hardly be seen by the naked eye. As in stems, larval excreta are then pushed out through the hole, which becomes apparent after several days of mining activity. Inside the tuber, the larva tunnels just under the surface at first, but later penetrates more deeply. GPTM larvae feed exclusively on tubers during potato cultivation and during storage. Damage is caused by larvae that bore galleries into the tubers. After the larvae have left tubers, the exit hole is clearly visible. In potato fields, *T. solanivora* attack occurs from tuberization until harvest (Kroschel and Schaub 2013; Niño 2004). Tubers infested by either of the three species develop a bitter taste and are unsuitable for human or livestock consumption (Keller 2003; Kroschel and Schaub 2013).

Impacts on production losses

P. operculella. Under heavy field infestation, potato foliage can be destroyed, which can result in substantial yield loss of up to 70%. High infestations early in the season can directly affect tuber yield. Strong correlation exists between leaf and consequent tuber infestation, which suggests that reducing *P. operculella* population density during the growing period is key to reducing potato tuber infestation at harvest. Hence, the most devastating yield losses are largely a result of earlier tuber infestation in the field, generally where moths have laid eggs through soil cracks on the developing tubers, or when harvest is delayed. *P. operculella* also damages harvested potato tubers in storage. The damage to potatoes in rustic stores can be total within a few months if the tubers are left untreated. Infested tubers are unsuitable not only for human consumption but also for use as seed. Infested tubers produce fewer yields and initiate a fast development of a new field *P. operculella* population (Kroschel and Schaub 2013; Kroschel et al. 2012; Keller 2003; Kroschel 1994, 1995).

S. tangolias. This species has become an economically important pest in potato fields and in storage in mid-elevation regions of the Andes (Peru, Bolivia, and Ecuador); its status in Colombia is not well confirmed. In the Andes, losses in the field may reach up to 30%, but most economically significant damage occurs when infested tubers are transferred to potato stores where reinfestation takes place. Without adequate management, farmers can completely lose their house-stored potatoes within 3–4 months of storage. In Australia and New Zealand, where the pest was accidentally introduced from South America, it is more recognized as a local pest of tomato, poroporo, sweet cucumber and other Solanaceae crops, and is commonly referred to as "tomato stem borer." However, at national level it is considered a minor pest and its economic impact on these crops is not well reported in the literature (Kroschel and Schaub 2013; Keller 2003).

T. solanivora. Complete losses of harvested tubers have been observed occasionally after the invasion of the pest into new areas when farmers were not yet familiar with pest control. Generally, tuber damage rates at harvest vary between 2 and 15% in the Andean region. When infested potato tubers are stored without application of control methods *T. solanivora* can destroy, depending on the storage period and temperature, of whole potato stock (Kroschel and Schaub 2013; Niño 2004).

Methods of prevention and control

Control of the potato tuber moths must take place both in the field and in storage. Implementation of integrated pest management is recommended to reduce the pest problem in field and stores. *T. solanivora* is the most difficult potato moth to control as the larvae feed only inside potato tubers where they are hard to reach (Kroschel and Schaub 2013; Kroschel et al. 2012; Keller 2003; Pollet et al. 2003; Kroschel 1995).

Monitoring with pheromone traps. For all three potato tuber moth species sexual pheromones have been identified and synthesized. They are used for monitoring the flight activity of adult male populations to detect early the presence of the different moths in the field and store to take adequate control measures.

Cultural practices. Some common practices for potato tuber moths are the use of pest-free seed tubers, deep planting, regular irrigation to avoid soil cracking, high hilling to protect tubers, timely harvest, not leaving the tubers after harvest exposed in the field for a long time (especially throughout the night), i.e., harvest and store immediately, and removal of leftover tubers to reduce the overwintering field population. Also, early maturing varieties can contribute to reduced risk of infestation.

Biological control. Classical biological control can be an effective strategy in all those regions in which the pests has been unintentionally introduced to keep the pest population below economic threshold; for this approach, the species *Copidosoma koehleri* (Blanchard), *Apanteles subandinus* (Blanchard), and *Orgilus lepidus* (Muesebeck) have been widely and successfully used (Canedo et al. 2016a, b, c; Kroschel and Schaub 2013).

Biopesticides. Microbial biopesticides for *P. operculella* field control have been tested based on *Bacillus thuringiensis* subsp. *kurstaki* (*Btk*) and *P. operculella*-specific granulovirus (*Phop*GV, Baculoviridae). *Btk* was effective but required repeated applications because it is quickly degraded by UV light. Likewise, *Phop*GV has shown mixed results. To protect *Phop*GV against UV inactivation a variety of adjuvants (e.g., dyes, optical brighteners) have been tested but simple preparations of *Phop*GV-infected larvae macerated in water were superior. Applications of *Phop*GV doses sufficient to cause >95% mortality are considered not being economical, and low dose treatments are proposed for a relatively inexpensive partial suppression of the field population (Lacey and Kroschel 2009; Sporleder and Kroschel 2008; Kroschel and Sporleder 2006; Sporleder 2003; Kroschel et al. 1996).

Attract-and-kill. This approach has been developed to control of *P. opercullela* and *S. tangolias* under field and storage conditions. It consists of a co-formulation of the insect pest-specific sexual pheromone, which "attracts" males, and a contact insecticide at very low concentration which "kills" males getting in contact with the product. The oil formulation is applied at a droplet size of 100 μL using a special handheld applicator; it is applied at 2500 droplets/ha. It effectively reduces the male population and the number of offspring, hence controlling larvae damage in the crop. It provides pest-specific control, and is harmless to natural enemies, humans, and the environment (Kroschel and Zegarra 2010, 2013). In Peru, the two products AdiosMacho-*Po*® and AdiosMacho-*St*® have been registered to be commercialized in Peru and the Andean region.

Chemical control. Broad-spectrum insecticides have been commonly used to suppress potato tuber moth population and economic damage, but which has been associated with many negative effects causing resistance of the pests to various active ingredients and affecting farmers and the environment.

Integrated Pest Management. Effective IPM practices for potato tuber moths have been developed, which can be applied successfully if potato tuber moths are the only economically important pests in an agroecosystem (e.g., Republic of Yemen, Kroschel 1995). However, potato is often affected by several pest species which requires a system approach to manage all economically important potato pests (Kroschel et al. 2012). This means effective IPM practices are required for all pests to fully eliminate or minimize the use of insecticides.

Storage management

Potato tuber moth infestation occurs frequently in rustic farmer-managed potato stores in developing countries, especially if temperature is suitable for rapid population build up and the storage lasts for several months. Storage facilities should be cleaned thoroughly before potato tubers are stored. Fine netting at windows should protect adult moths from entering storage facilities. Only healthy tubers should be selected for storage. Infested potato tubers need to be destroyed. However, initial infestations cannot be easily observed and are the main reasons why potato tuber moths enter storage facilities and infest potato. Sex-pheromone-baited water traps or funnel traps (Delta) can be used for monitoring the moth presence but in known region of potato tuber moths' occurrences, potatoes should be treated before storage.

Biopesticides. Biopesticides based on *Btk* and *Phop*GV are used in potato tuber moth storage control. The microbials are formulated in inert materials (e.g., talcum) and dusted over potatoes before storage. Since *Phop*GV is only effective in *P. operculella* and *T. solanivora,* in regions where all three species occur simultaneously, the use of *Btk* has the advantage to control all three species. Further, *Btk* is mostly available as a commercial biopesticide while *Phop*GV has to be multiplied in potato tuber moth larvae. The product Matapol, e.g., is a co-formulation between *Btk* and *Phop*GV, commercialized in Bolivia. It has also been shown that inert materials (e.g., calcium carbonate, kaolin, talcum, silicium rich sand) can be used effectively without the addition of active biologicals (*Btk,* or *Phop*GV) as they control first instar larvae through desiccation (Kroschel and Koch 1996; Mamani et al. 2011; Sporleder and Lacey 2013; Schaub and Kroschel 2017).

Attract-and-kill. Attract-and-kill formulations (see above) can be applied at a density of one drop (100 μL)/qm of storage area to reduce the male population and hence tuber infestation in potato stores (Kroschel and Zegarra 2013).

Chemical control. Malathion dust (WHO Class III) is often observed to be sold in developing countries to treat stored tubers. This is especially critical if precautions are not taken properly by farmers and potatoes are stored in living areas. Pyrethroids (e.g., fenvalerate) have shown to be highly effective equally to *Btk* treatments described above (Kroschel and Koch 1996).

Pea Leafminer Fly

Liriomyza huidobrensis Blanchard (Diptera: Agromyzidae)

Distribution The pea leafminer *Liriomyza huidobrensis* is an agricultural pest endemic to South America. Since the early 1980s, the pest has been also recorded in many other countries around the world, presumably associated with the global trade of ornamental plants (Mujica et al. 2016; CABI 2012).

Fig. Adult of *Liriomyza huidobrensis* (**a**) and symptoms of the larvae infestation on potato leaves (**b**). (Photo credits: CIP)

Host range *L. huidobrensis* is highly polyphagous and has been recorded from plants of 14 families (Spencer 1973, 1990). The long list includes potato, bean, pea, alfalfa (*Medicago sativa* L.), and vegetables such as tomato, celery (*Apium graveolens* var. dulce (Mill.) Pers.), lettuce (*Lactuca sativa* L.), pepper (*Capsicum annuum* var. *longum* (DC.) Sendtn.), and spinach (*Spinacia oleracea* L.). In addition, leafminer flies infest many weed species and ornamental plants.

Symptoms of infestation Adults and larvae of *L. huidobrensis* damage the plant foliage. Adults cause damage by puncturing the leaf surface to feed on the leaf tissue, and to lay eggs. Newly hatched larvae mine into the leaf and feed on the chloroplast-rich mesophyll, making a serpentine mine whose diameter increases as the larva grows. A large proportion of grown larvae remain close to the midrib. Leaf tissue affected by larval mining becomes necrotic and brownish. Highly infested crop fields appear burned (Cisneros and Mujica 1999b).

Impacts on production losses *L. huidobrensis* is a serious pest of arable crops, vegetables, and ornamental plants under field and glasshouse conditions in many parts of the world. Plant injuries caused by adult and larval activities reduce photosynthesis activity and cause leaf wilting. For potato, yield losses of up to 100% were reported in Argentina, Chile, and Indonesia (Cisneros and Mujica 1999b). Pest intensity-crop loss relationships for the leafminer fly in different potato varieties indicated that the accumulated foliar injury up to the growth stages of flowering and berry formation produced the highest yield losses in the different potato varieties (Mujica and Kroschel 2013). Economic injury levels in Peru varied according to control costs and commodity values, and potato varieties with longer vegetation period can tolerate higher levels of foliar injury by the leafminer fly before control measures are needed (Desiree: 21–28%, Revolucion: 34–47%, Canchan: 31–40%, Maria Tambeña: 40–53%, Tomasa: 55–74%, and Yungay: 40–54% of foliar injury).

Methods of prevention and control

Ecological and economical sound control of the leafminer fly is best realized when based on IPM, which promotes natural enemies in combination with cultural practices and low-toxic insecticides (Mujica 2016; Mujica et al. 2016; Kroschel et al. 2012; Weintraub et al. 2017).

Monitoring pest populations. Counting the number of flies captured in yellow sticky traps monitors adult leafminer fly activity. Counting the number of larvae or fresh tunnels per leaflet by sampling the bottom, middle, and top parts of the plant is used to monitor larval infestation. Both methods can be adapted for decision making and applying an action threshold (AT) to avoid unnecessary applications of insecticides. The AT can be defined as the level of pest population at which control measures should start to prevent the pest population from reaching an economic injury level (EIL, point where economic losses will begin). The AT is typically set below the EIL accounting for the lag time to implement effective control measures. Preemptive insecticidal control is economically not justified until foliar injury exceeds these values (Mujica and Kroschel 2013; Mujica and Kroschel 2018).

Crop management. Healthy, vigorous growing potato plants can better tolerate leafminer damage, particularly during the vegetative phase. Balanced N-fertilization is important as high N-content in leaves promotes leafminer fly development. Continuous food availability by replanting host crops will favor the abundance of the leafminer fly. Rotation with nonhosts is therefore recommended.

Conserving beneficial insects. Leafminer flies are controlled by many beneficial insects, which are either predators or parasitoids. Strategies to conserve beneficial insects can be manifold and include diversified cropping systems, high structural floristic diversity in agricultural landscapes, special weed management practices, and reduced use of broad spectrum insecticides (Mujica et al. 2016).

Classical biological control. It can be an effective strategy in all those regions in which the pea leafminer fly has been unintentionally introduced and where natural enemies of leafminer flies are absent to keep the pest population below economic threshold. The endoparasitoids *Halticoptera arduine* Walker (Pteromalidae), *Chrysocharis flacilla* Walker (Eulophidae) and *Phaedrotoma scabriventris* (Nixon) (Braconidae) were successfully introduced and established in three agro-ecological regions (low, middle, and high altitude) in Kenya (Muchemi et al. 2014; Mujica et al. 2016). Yet, biocontrol of leafminer flies needs to be accompanied by additional, biocontrol-compatible control measures which need to be ultimately integrated into one holistic IPM concept for vegetables that addresses all major pests of the system.

Use of entomopathogenic nematodes. The entomopathogenic nematode *Heterorhabditis indica* (Rhabditida: Heterorhabditidae) caused 58.7% of leafminer larval mortality in potato leaves under semi-field conditions. It could be considered as biocontrol-compatible control measure as part of an IPM program for potato and vegetables (Mujica et al. 2013).

Physical control. Yellow attracts leafminer fly adults. The use of mobile (1 × 4 m bands of stick traps which are moved across the fields fixed at tractors or hand-carried) and stationary (50 × 50 cm, 60–80 traps/ha) yellow sticky traps can effectively reduce the leafminer fly adult population. In the Cañete valley of Peru, a cumulative capture of up to seven million adults/ha by using fixed and mobile yellow sticky traps (US$66.7/ha) resulted in a reduction of the control costs by 55.5% compared with chemical control ($200.0/ha), and an average use of six adulticide applications per season (Mujica et al. 2000).

Chemical control. Decisions to use insecticides should be made according to the monitoring results and when the leafminer population is expected to cause economic damage (Mujica and Kroschel 2013). Systemic insecticides with translaminar properties are most effective in controlling leafminer fly larvae. Such insecticides include abamectin and spinosad or cyromazine (Weintraub 2001).

Integrated Pest Management. An IPM strategy based on the use of seed treatment, action threshold, trapping devices and selective application of insecticides showed a higher efficacy to control potato pests including *L. huidobrensis* than the conventional application of insecticides by farmers in the Cañete valley of Peru. IPM reduced the total quantity of pesticides used per season by 56% compared to the conventional management, representing a decrease of 69.2% in the environmental impact. Further, IPM achieved 35% of higher marketable potato yield than conventional management (Mujica and Kroschel 2018). Leafminer fly is a polyphagous pest and biocontrol-compatible control measures need to be integrated into an IPM concept which considers also other economic pests in potato.

Andean Potato Weevils

Premnotrypes suturicallus **Kuschel, *P. vorax*** **(Hustache), *P. latithorax*** **(Pierce) (Coleoptera: Curculionidae)**

The Andean potato weevil complex consists of at least 14 species with 12 in the genus *Premnotrypes* and two in the genus *Rhigopsidius* and *Phyrdenus*. The most important species attacking potato are *Premnotrypes vorax*, *P. latithorax* and *P. suturicallus.* All *Premnotrypes* species show sexual dimorphism; the females (6.8–8.0 mm) are larger than males (5.6–7.5 mm) (Alcázar and Cisneros 1999).

Distribution The Andean potato weevils (*Premnotrypes* spp.) are the most serious pests of the potato in the Andean region above 2800 m above sea level. Its distribution extends from Argentina to Venezuela, covering a mountainous territory of about 5000 km in length. These weevils are native to the Andes where wild potato and cultivated species are their hosts (Alcázar and Kroschel 2008).

Host range The host range of Andean potato weevils includes only potato and its wild relatives.

Symptoms of infestation The Andean potato weevil, in the larval and adult stages, causes damage to the potato. Adults feed on leaves starting from their edges, marking a very characteristic form of semicircle. When the beetle population is very

Fig. Adult of *Premnotrypes suturicallus* (**a**: male, left; female, right), and symptoms of weevil damage on leaves (**a, b**) and of larvae infestation in potato tubers (**c**). (Photo credits: CIP)

high, leaves are eaten up to the central leaf vein. Occasionally, adults can also damage on stolons, tubers during formation, and the stem bases. Serious damage to tubers begins when newly emerged larvae penetrate tubers. As the larvae develop, they build characteristic tunnels that are usually filled with excreta. Once the larval stage is complete, the larvae leave the tuber making characteristic circular exit holes (Alcázar and Cisneros 1999; Alcázar and Kroschel 2008).

Impacts on production (losses) Andean potato weevils cause substantial yield losses that seriously threaten Andean farmers' food security. Losses largely vary (16–45%) even when insecticides are applied. If weevils are not routinely controlled, losses can even reach 80–100% (Ortiz et al. 1996; Kroschel et al. 2012).

Methods of prevention and control
(Alcázar and Cisneros 1999; Alcázar and Kroschel 2008; Kroschel et al. 2012).

Cultural control. The main sources of weevil infestations are potato fields of the previous season. It is recommended to maintain community rotation system over distances of about 1 km between fields. Let the field rest for 3–5 years before planting potato and avoid planting potatoes for two consecutive seasons. Best candidate crops to rotate with include faba bean and barley. Elimination of volunteer potato plants in other crops should be practiced. Early harvest and elimination of crop residues using animals (pigs, sheep) reduces weevil populations in the field.

Mechanical control. Night collection of adults through plant shaking into buckets.

Physical barriers. It is a method that prevents the adult weevils from entering the field by using a plastic barrier placed around the potato field. It is recommended to install the barriers before or at the time of planting, to prevent the first adults from entering the new fields. Results from the use of plastic barriers studies revealed a higher efficacy for the control of the Andean potato weevil than applications of insecticides (Kroschel et al. 2009).

Plant barriers. Interrupting adult migration into potato fields through plant barriers of tarwi (*Lupinus mutabilis* Sweet) or mashua (*Tropaeolum tuberosum* Ruíz and Pavón) has also shown effects of reducing the weevil population.

Chemical control. Generally, the use of insecticides has not demonstrated a consistent efficacy and farmers still experience losses while spraying insecticides 2–3 times.

Natural enemies. Although the weevils are native to the Andes, no specific parasitoids have been identified but predators like carabids are wide-spread and affect the weevil population. Most common species are *Blennnidus* sp., *Notiobia schnusei* (Van Endem) and *Harpalus turmalinus* Er. In addition, entomopathogenic fungi (*Beauveria bassiana* (Balsamo) Vuillemin) and nematodes (*Heterorhabditis* sp., and *Steinernema* sp.) have been identified and used to develop biocontrol strategies (Kroschel et al. 2012; Alcázar and Cisneros 1999; Kaya et al. 2009).

Integrated Pest Management. An IPM strategy consists of the use of plastic barriers in combination with cultural practices; one insecticide application might be required if plastic barriers are used in potato–potato rotations (Kroschel et al. 2011, 2012).

Potato Psyllid

Bactericera (ex-Paratrioza) cockerelli **(Sulc) (Hemiptera: Triozidae)**

Distribution The potato psyllid *Bactericera cockerelli* is native to North America and occurs mainly in the Great Plains region of the United States, from Colorado, New Mexico, Arizona, and Nevada, north to Utah. It is also found in Mexico, Guatemala, Honduras, and Nicaragua, and is suspected to be present in other Central American countries, including El Salvador. Following an accidental introduction, *B. cockerelli* has become widespread in New Zealand (CABI 2016; OIRSA 2016; Rehman et al. 2010).

Host range *B. cockerelli* is found primarily on plants of the family Solanaceae. The psyllid infests and develops on a variety of cultivated and weedy plant species including potato, tomato, pepper, eggplant, and tobacco as well as noncrop species such as black nightshade, groundcherry (*Physalis* spp.), and matrimony vine (*Lycium* spp.) (CABI 2016).

Fig. Nymphs of *Bactericera cockerelli* infesting potato leaves (**a**) and potato plant affected with potato purple top (**b**). (Credits: Courtesy of CIP)

Symptoms of infestation Both adults and nymphs feed by sucking the sap of plants. Direct damage is caused by injecting substances that destroy plant cells (toxins). Chlorophyll production is interfered causing leaf yellowing and stuntedness, which together result in a condition known as psyllid yellows. Indirect damage is related to the transmission by nymphs and adults of the bacterium *Candidatus liberibacter solanacearum* (psyllaurous) that causes infectious diseases known as permanent tomato (PT), potato purple top (PM), and zebra chip. Typical symptoms of potato zebra chip include yellowing and curling of foliage, stunted growth, formation of aerial tubers, shortened and thickened internodes, leaf scorching, reduced tuber size and yield, and early plant death. Belowground, zebra chip is characterized by the presence of collapsed and necrotic stolons, and browning of internal vascular tissues, which, upon frying, exhibits dark brown streaks, hence the term "zebra chip" (Martin 2016; Rehman et al. 2010; Garzón 2002).

Impacts on production losses Potato zebra chip disease is a serious disorder of potatoes that has resulted in millions of dollars in losses to the potato industry. The characteristic symptom is dark striping in the tuber when fried as French fries or potato chips, rendering the infected potatoes unmarketable. It has been determined that *C. liberibacter* is transmitted to potato very rapidly by the potato psyllid, and that a single psyllid per plant can successfully transmit this bacterium to potato in as little as 6 h. This low psyllid density, coupled with a short inoculation access period, represents a substantial challenge for growers in controlling the potato psyllid and preventing zebra chip transmission. Just a few infective psyllids feeding on potato for a short period could result in substantial spread of the disease within a potato field or region (Garzón 2002; Rehman et al. 2010; Munyaneza et al. 2007, 2009).

Methods of prevention and control

Monitoring pest population. Early season management of this insect is crucial to minimize damage and psyllid reproduction in the field. Egg and nymphal stages require visual examination of the foliage. Adults can be sampled with yellow traps, which are effective in detecting immigrant populations and must be installed from the beginning of the cropping season. Low populations before or at the start of tuber formation reduce production significantly, but once the tubers are formed plants tolerate direct damage (Henne et al. 2010; Butler and Trumble 2012; Workneh et al. 2012).

Plant resistance. The most valuable and effective strategies to manage zebra chip would likely be those that discourage vector feeding, such as use of plants that are resistant to psyllid feeding or less preferred by the psyllid. Unfortunately, no potato variety has so far been shown to exhibit sufficient resistance or tolerance to zebra chip or potato psyllid.

Biological control. Several predators and parasites of *B. cockerelli* are known, though there is little documentation of their effectiveness. In New Zealand, fungal isolates of *Lecanicillium muscarium* (Petch) Zare & W., *Isaria fumosorosea* (Wize), and commercial formulations of *Beauveria bassiana* Balsamo-Crivelli (Vuillemin), *I. fumosorosea*, and *Metarhizium anisopliae* (Metchnikoff) Sorokin were successfully tested against nymphs of *B. cockerelli* in laboratory assays and adults and nymphs in greenhouse assays, outperforming a conventional spiromesifen insecticide (Mauchline and Stannard 2013). Commercial formulations of *Metarhizium anisopliae* (F52®, Novozymes Biologicals) and *Isaria fumosorosea* (Pfr 97®, Certis USA) significantly reduced plant damage and potato zebra chip symptoms in Weslaco, Texas (Lacey et al. 2011).

Chemical control. Good insecticide coverage or translaminar activity is important because psyllids are commonly found on the underside of the leaves. Because several generations often overlap, caution should be taken when selecting and applying insecticides targeted against the potato psyllid in relation to which life stages are present in the crop and timing of insecticide applications.

Integrated Pest Management. B. cockerelli is often associated with the pathogen *Candidatus liberibacter solanacearum* and this represents the main challenge in the search for IPM components that can be integrated into the potato crop. Several management strategies have been proposed such as cultural, biological, and chemical control. In New Zealand, the proposed IPM strategy includes careful timing of insecticide applications, incorporating alternative chemicals, integrating treatments with vector monitoring or sampling, establishing action thresholds (although this is difficult for vectors), determining the likelihood of infections throughout the year, and combining chemical applications with resistant or tolerant varieties (Vereijssen et al. 2018).

Bud Midge

Prodiplosis longifila Gagne (Diptera: Cecidomyiidae)

Distribution *Prodiplosis longifila* is a polyphagous species only in the Americas Its origin is unknown. It occurs in North America (first reports from Florida in 1934 and from Virginia in 1990) and South America (Colombia, Peru, and Ecuador). Some publications mention its presence in the "West Indies" but detailed records for individual countries in the Caribbean are not available and a previous record for Jamaica is now considered to be a misidentification (EPPO 2015; Gagné 1986).

Host range *P. longifila* is polyphagous and has been recorded from many plant species. However, economic damage is mainly reported on potato, tomato, pepper, asparagus and to a lesser extent on Tahiti lime (*Citrus* × *aurantifolia* (Christm.) Swingle, *Citrus* x *latifolia* (Yu. Tanaka) Tanaka). Lists of host plants include important crops such as onion (*Allium cepa* L.), watermelon (*Citrullus lanatus* (Thunb.) Matsum. & Nakai), melon (*Cucumis melo* L.), cucumber (*Cucumis sativum* L.), artichoke (*Cynara scolymus* var. *scolymus* L.), soybean (*Glycine max* L.), alfalfa (*Medicago sativa* L.), beans (*Phaseolus vulgaris* L.), castor bean (*Ricinus communis* L.), and grape (*Vitis vinifera* L.). Because the identity of the different populations of *P. longifila* has not yet been resolved, some records of hosts previously attributed to *P. longifila* may belong to another species (EPPO 2015).

Symptoms of infestation *P. longifila* is minute in size, but extremely injurious under warm environmental conditions. The larva passes through three stages, with the first and second larval stages being the most harmful and causing damage to flowers and buds of different crops. The larvae of *P. longifila* scrape the epidermal tissues of plant structures using piercing-sucking mouthparts. Feeding on buds distorts growth points and results in winding stems and black appearance. It causes

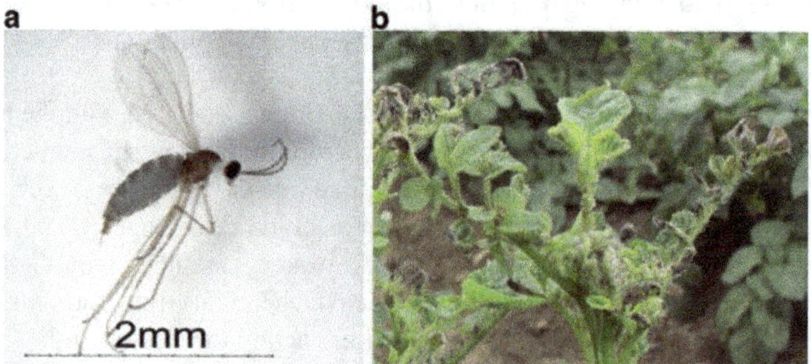

Fig. *Prodiplosis longifila* adult (**a**), larvae (**b**) and damage on potato sprouts (**b**). (Photo credits: CIP)

flower buds to fall and results in distorted fruit of low quality (EPPO 2015; Hernandez et al. 2015; Sarmiento 1997).

Impacts on production losses *P. longifila* is one of the most important pests of Solanaceae crops (potatoes and other vegetables) and asparagus in the Neotropics causing yield losses >50%, if not controlled. It affects buds, flowers, and fruits, and causes severe damage to both open and protected crops. During the first 50 days after plant emergence, the bud midge is a critical pest that affects plant growth by destroying both the terminal and lateral buds. However, infestations after flower initiation do not affect potato yields. High temperatures favor the increase in midge population and hence crop damage, which could explain the reduced potato yields recorded during the El Niño phenomenon in Peru (EPPO 2015; Hernandez et al. 2015; Sarmiento 1997).

Methods of prevention and control

Crop management. Eggs and first instar larvae are in protected places (e.g. under the calyx) and are not visible with the naked eye (Gonzales-Bustamante 1996). Therefore, factors that favor vigorous potato plant growth (appropriate fertilization, healthy seed with vigorous sprouts, irrigation, and biostimulants) prevent pest damage by the rapid growth of buds and the exposure of larvae to dehydration and natural enemies. Management practices to reduce infestation include optimal planting density, orientation of the furrows from east to west to favor insolation, elimination of broadleaf weeds, avoidance of susceptible crops, management of the planting season, and rotation with nonsusceptible crops (Mujica and Kroschel 2018).

Seed treatment. Seed treatment with a systemic insecticide protects buds against initial bud midge infestation from emergence to formation of the first leaves and stems. This is a highly recommended practice since it ensures protection for a period of 35 days during a critical period of the crop, without affecting natural enemies (Mujica and Kroschel 2018).

Chemical control. For a chemical control of this pest, the following should be considered: (a) dusting of sulfur powder on young plants to reduce initial infestation, (b) the use of translaminar or systemic insecticides, (c) dusting of dry powder insecticides at the soil or lower third of the plants, and (d) a combined control of larvae and adults with various active ingredients such as imidacloprid, dimethoate, mixtures of chlorpyrifos and alphacypermethrin, and vegetable and mineral oils (Castillo 2010). Sulfur dust has repellent effects on the adults, but low toxicity to larvae. Treatments with sulfur are more effective in the late afternoon when adults become active and more exposed to the treatments.

Biological control. The most commonly occurring biological control agents are egg and larva parasitoids of the genus *Synopeas* (Hym.: Platygasteridae), reaching parasitism of 16–20% (Hernández 2014). Important predators of third instar larvae and pupa are *Chrysoperla asoralis* Banks, *Nabis capsiformis* Germar, and *Methacantus tenellus* Stal.

Biopesticides. In Peru, the entomopathogenic nematode *Heterorhabditis* sp. (native) showed 80% pathogenicity in larvae and pupae of *P. longifila* (Pacheco 2015). Mortality of *P. longifila* caused by *Metarrhizium anisopliae*, *Baeuveria bassiana*, *Paecilomices fumosoroseus*, and *Verticillium lecanii* under laboratory and greenhouse conditions was 75 and 69%, 71 and 73%, 62 and 50%, 59 and 44%, respectively (Reategui et al. 2009). The application of entomopathogens in commercial fields of asparagus showed a reduction of *P. longilifa* pupae by 79.2% (*B. bassiana*), 70.8% (*M. anisopliae*), 30.4% (*V. lecanii*) (Castillo 2010).

Major Pests in Temperate Regions

Colorado Potato Beetle

***Leptinotarsa decemlineata* (Say) (Coleoptera: Chrysomelidae)**

Distribution The Colorado potato beetle *Leptinotarsa decemlineta* is native to Mexico, but has spread to Central America, most of the United States (except Alaska, California, Hawaii, and Nevada), and to southern Canada. This pest has also been introduced into Europe and parts of Asia (CABI 2017a; Jacques and Fasulo 2015).

Host range Colorado potato beetle is an oligophagous species feeding on about ten species of plants in the family Solanaceae (Alyokhin et al. 2013). These include potato, tomato, and eggplant.

Symptoms of infestation Both adults and larvae are voracious leaf feeders and can completely defoliate potato plants. In rare cases, tubers exposed at the soil surface can also be eaten. Characteristic black and sticky excreta is left on the stem and leaves by the larvae and adults.

Impacts on production losses *L. decemlineata* is one of the most economically important insect pests of potato. Adults and larvae reduce potato tuber yields by

Fig. Colorado potato beetle, Leptinotarsa decemlineata, adults (**a**) and larvae feeding on potato leaves (**b**). (Photo credits: Andrei Alyokhin)

devouring foliage. Larvae may defoliate potato plants resulting in yield losses of up to 100% if the damage occurs prior to tuber formation. Losses in potato are most severe if the incidence of last instar larvae, the most voracious stage, peaks during the time of tuber formation. This species is a huge pest problem and farmers spend millions of dollars each year on insecticides to control it (CABI 2017a; Jacques and Fasulo 2015).

Methods of prevention and control

IPM programs against this potato pest emphasize reducing insect pest populations by using several control measures, including crop rotation, altered planting dates (to avoid peak pest populations during vulnerable stages of crop development), and use of *B. thuringiensis* toxins or other pesticides when economic injury levels are reached. IPM approaches seek to maximise the impact of naturally occurring biological control agents in suppressing beetle populations (CABI 2017a; Jacques and Fasulo 2015; Laznik et al. 2010; Capinera 2001; Ferro 2000).

Sampling. Entire plants should be examined for above-ground life stages of the insect. Adult beetles are attracted to the yellow color and can be captured with traps. However, traps are seldom used because the various life stages are clearly distinguishable from each other.

Cultural control. The Colorado potato beetle may be managed culturally by crop rotation. Distances of at least 0.5 km between potato fields are required to provide protection. Beetles initially disperse by walking, so crop rotation and/or trenching can significantly reduce field infestations. Destruction of crop debris after harvest may contribute to reducing beetle populations. Beetle densities are also often lower on mulched potato plots (Alyokhin 2009).

Biological control. While many natural enemies have been identified, they are usually not able to control Colorado potato beetle populations below the economic injury levels. The tachinid fly *Myiopharus doryphorae* (Riley) may produce high parasitism rates early in the season and prevent the beetles from becoming a serious pest. Also, the entomopathogenic nematodes (*Steinernema feltia* Filipjev) and the fungus *Beauveria bassiana* have been reported to offer some control against this pest (Alyokhin 2009).

Biopesticides. Abamectins and spinosins generally provide excellent control of this pest (but see cautionary note on insecticide resistance below). Bacterial insecticides based on delta endotoxin of bacterium *Bacillus thuringiensis* subsp. *tenebrionis* are also effective, but they must be applied against the first two instars.

Chemical control. Insecticides are commonly used to control populations of Colorado potato beetle. Resistance to insecticides develops very rapidly in this species. For example, for endrin the first failure was reported after 1 year of its use, while for oxamyl it was observed already within the first year of use (Forgash 1985). Overall, failures were reported to 51 compounds belonging to ten chemical groups and possessing eight modes of action (Alyokhin et al. 2013).

European Corn Borer

Ostrinia nubilalis (Hübner) (Lepidoptera: Crambidae)

Distribution *Ostrinia nubilalis* is native to Europe and was introduced into North America near Boston (Massachusetts) in 1917 (Baker et al. 1949). It gradually spread from there to other parts of the United States and Canada. It has also been reported in Asia (China, India, Georgia and Indonesia, middle East (Syria, Israel, Lebanon, Iran, Turkey) and North Africa (Algeria, Egypt, Libya, Morocco and Tunisia) (EPPO 2014; Li et al. 2003). Two races, E and Z that are morphologically indistinguishable exist in both USA and Europe.

Host range The European corn borer is primarily a pest of maize (*Zea mays* L.), but very polyphagus feeding on host plants from over 17 families but not necessarily being a pest. Economic host crops include potato, pepper, celery (*Apium graveolens* L.), tomato, beans, hop (*Humulus lupulus* L.), oat (*Avena sativa* L.), millet (*Panicum* spp.), sorghum (*Sorghum bicolor* L. Moench), cotton (*Gossypium arboreum* L.), and fruits such as apple (*Malus pumilla* Miller) and peach (*Prunus persica* (L.) Batsch L.). Feeding also occurs in ornamental plants and weeds including hollyhock (*Alcea rosea* L.), mugwort (*Artemisia vulgaris* L.), pigweed (*Amaranthus* spp.), and others.

Symptoms of infestation In potato, the most obvious sign of infestation is stem wilt. The entrance hole is in the stem and is easy to locate because of the presence of the excreta that larvae expel while feeding inside the stem. Larval presence can be confirmed by cutting the stem (CABI 2017b). Young larvae tunnel into leaf petioles while older larvae tunnel into the main stem. Larval damage can, however, be confused with that of other stem borers.

Impacts on production losses Potato plants present high tolerance for this insect; fairly high levels of infestation often do not seem to significantly affect tuber yields (Kennedy 1983). Varieties with weak stems are more likely to suffer reduced yields, as affected stems will break more easily during heavy wind- or rain- storms. However, severe infestations may result in considerable crop losses. Also, larval entry holes may facilitate infection by fungal and bacterial pathogens such as *Erwinia carotovora*.

Methods of prevention and control
Sampling. Monitoring of adult moth flights is suggested to determine the timing for oviposition and hence if a field is at risk. Sticky traps baited with strain-specific sex pheromones can be used for this purpose. Early infestations can also be more damaging than later ones and scouting for egg masses and observation of damage by the first instar larvae on leaf petioles. It is prudent to be vigilant and to monitor populations in areas where this insect has been known to previously cause damage.

Cultural control. This can involve early planting to escape damage, intercropping with nonhost plants and proper field sanitation involving of ploughing the field at the end of the cropping season.

Chemical control. European corn borer larvae are not particularly difficult to kill, but pesticide applications, to be effective, must be made before the larvae enter the stem. Pesticide application is recommended only when more than 10% of the stems show evidence of tunneling (Nault and Kennedy 1996).

Biological control. *Bacillus thuringiensis* subsp. *kurstaki* (*Btk*) and *B. thuringiensis aizawai* can be successfully applied to target first instar larvae. Alternative control methods include release of commercially available parasitic wasps (*Trichogramma nubilale* Ertle and Davis and *T. brassicae* Bezdenko) (CABI 2017b). Other natural enemies that have been used to control the European corn borer include *Lydella thompsoni* (Herting), *Diadegma terebrans* (Gravenhorst) and *Macrocentrus grandi* Goidanich (Baker et al. 1949).

Major Pests Globally Present

Aphids

Myzus persicae (Sulzer, 1776), *Macrosiphum euphorbiae* (Thomas, 1878) (Hemiptera: Aphididae)

Distribution The green peach aphid, *Myzus persicae*, is thought to have its origins in China, just as its overwintering host plant, the peach (*Prunus persica* (L.) Batsch) However, the green peach aphid is highly adaptable and is currently cosmopolitan in distribution. The potato aphid (*M. euphorbiae*) originated in North America but has spread to the temperate parts of Europe and Asia and is found in all potato growing areas globally (CABI 2017c, d).

Host range Both species are extremely polyphagous, being capable of feeding on several hundred-plant species (Blackman and Eastop 2000).

Symptoms of infestation Aphids can damage potato plants directly by feeding on sap, and indirectly by transmitting various viral diseases. *Direct damage*. Continuous

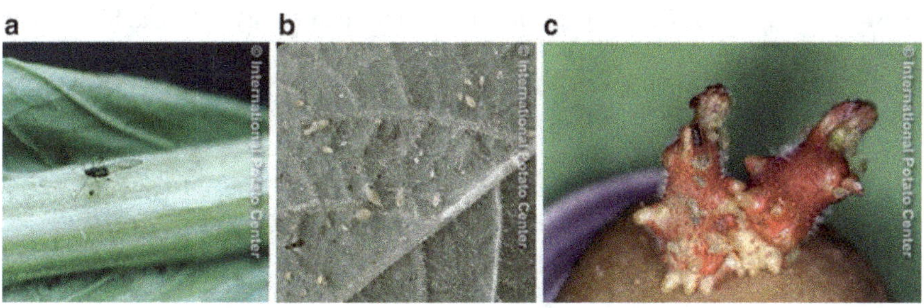

Fig. Winged green peach aphid start infestations on potato (**a**), wingless green peach aphids start colonies under the leaf (**b**), aphid colonies on potato tuber sprouts in stores (**c**), where they transmit viruses to seed potatoes. (Photo credits: CIP)

sucking of sap by large numbers of aphids considerably weakens and slows plant development. Removing sap and injecting toxic saliva causes leaf deformation. The weakened plant produces low quality tubers. During feeding, aphids excrete honey dew which promotes growth of a sooty black mold on plant leaves or stems, hence reducing the photosynthetic area for the plant. *Indirect damage*. The most important damage caused by aphids in potato is virus transmission. Symptoms of virus infestation vary depending on the virus transmitted. For instance, potato leaf roll virus (PLRV) causes leaf-rolling and tuber stem necrosis in potato. Other PLRV symptoms include thickening, curling, chlorotic spotting and yellowing of the leaves (CABI 2017c, d; Larrain et al. 2003; Salazar 1995).

Impacts on production losses *Direct damage*: Large build-up in aphid population may result in death of heavily infested plants. Since aphid distribution within potato fields is usually clumped, this results in patches of dead plants, sometimes referred to as "aphid holes." However, this phenomenon is relatively uncommon. In most cases, direct damage by aphids does not affect potato yields. *Indirect damage*. Most aphid-related damage to potato crops is caused by virus transmission. Potato leafroll virus (PLRV) and potato virus Y (PVY) are the two most important potato-infecting viruses. Both severely diminish potato yield and quality in most potato-growing areas. PLRV is a persistent virus transmitted exclusively by potato colonizing aphid species. PVY is nonpersistently transmitted by at least 50 different aphid species, most of which probe potato plants with their proboscis, but do not settle nor reproduce on them. *M. persicae* is the most effective vector of both persistent and nonpersistent viruses. It is also a highly polyphagous species that often moves into potato fields from surrounding crops and noncrop vegetation (CABI 2017c, d; CIP 1996; Larrain et al. 2003; Salazar 1995).

Methods of prevention and control

Biological control. Various natural enemies of aphids act as efficient biological control agents; among those are different predatory insects (e.g., lady beetles and larvae of green lacewings) and parasitic wasps (*Aphidius* spp.). Other generalist predators include *Orius* spp., *Geocoris* spp., and *Nabis* spp. Guard rows of flowering vegetation planted within potato fields can provide a habitant for aphid natural enemies (Powell and Pell 2017). Entomopathogenic fungi commonly infect aphids under field conditions and may cause epizootics that significantly reduce aphid densities (Pell et al. 2001). Numerous biological control products that use seven species of entompathogenic fungi (mostly *Beauveria bassiana* and *Lecanicillium* spp.) are commercially available for aphid control. Proper timing is very important when using these products because fungal spores are strongly influenced by environmental conditions, such as temperature and relative humidity (Kim et al. 2013).

Biological insecticides. Extracts from garlic (*Allium sativa* L.), neem (*Azadirachta indica* A. Juss.), red chili (*Capsicum annum* L.), pyrethrum flowers (*Chrysanthemum* sp.), either singly or in mixtures provide some aphid control, especially at an early stage of infestation.

Cultural practices. Weeding and removal of alternative and overwintering hosts such as wild mustards (*Brassica* spp.), use of wheat straw or white plastic as mulch, and intercropping with onion, garlic or coriander (*Coriandrum sativum* L.) have shown to reduce aphid populations. Planting only high-quality disease-free potato tubers and rogueing out virus-infected plants is recommended for preventing virus damage even in the presence of substantial aphid populations (CIP 1996; Larrain et al. 2003; Sanchez and Vergara 2002).

Chemical control. Insecticides specific to the homopterans such as spirotetramat, flonicamid and pymetrozine have low effect on natural enemies and are therefore good candidates for IPM. To the contrary, use of broad-spectrum insecticides may flair up populations of green peach aphids that recover quicker than populations of their natural enemies. On a small scale, application of potassium-based liquid soap sprays may reduce numbers of potato-colonizing aphids. When practicing chemical control, it is important to remember that green peach aphid has a high propensity for developing insecticide resistance that rivals that of the Colorado potato beetle. Failures have been reported for at least 69 different active ingredients, including all commonly used chemical classes (Alyokhin et al. 2013).

Whiteflies

***Bemisia tabaci* (Gennadius 1989), *Trialeurodes vaporariorum* (Westwood 1856) (Homoptera: Aleyrodidae)**

Distribution *Bemisia tabaci* is a common pest in tropical and subtropical regions but is less prominent in temperate habitats. *Trialeurodes vaporariorum* originated in tropical or subtropical America (probably Brazil or Mexico). Currently, it has become a cosmopolitan pest, with records in every zoogeographic region of the world (CABI 2017e, f).

Host range *B. tabaci* is a highly polypaghous pest. In addition to Solanaceae crops such as potato, tomato, and eggplant, it infests a wide range of other field crops and vegetables; among those are sweetpotato (*Ipomoea batatas* (L.) Lam.), cotton, beans, cucumber, melon, soybean, cassava (*Manihot esculenta* Crantz), and many others.

Symptoms of infestation Adult whiteflies can be easily seen on the underside of the leaves. Sucking sap from plant tissue is associated with several physiological plant disorders, such as chlorosis of new foliage. Heavy infestations with hundreds of adult greenhouse whiteflies on the lower surfaces of potato plant leaflets produce whitish spots that subsequently turn dark. As the population increases, sooty mold accumulates on the leaf surface, necrosis develops on the leaf margins, and some leaves curl upward. Necrotic leaf areas expand and coalesce, causing the whole leaf to dry-out. Producing sugar-rich honeydew promotes growth of sooty mold fungi, thus interfering with normal photosynthetic processes. Adult whiteflies transmit a number of viruses, each of which has its own suite of symptoms. For example, potato apical leaf curl virus (PALCV) causes upward or downward curling of leaves. Infection with potato yellow vein virus (PYVV) initially results in bright yellow

Fig. *Bemisia tabaci* adults on potato leaves (**a, b**) and potato plant with symptoms of Potato yellow vein virus (PYVV) transmitted by *Trialeurodes vaporariorum* (**c**). (Photo credits: CIP)

veins. As disease progresses, the leaves become yellow, while the veins regain their green color. Infected plants may produce deformed tubers with abnormally large protruding eyes (CABI 2017e, f; Larrain et al. 2003; Sanchez and Vergara 2002).

Impacts on production losses Both direct and indirect damage may result in yield reduction of 40% or more (Anderson et al. 2005; CABI 2017e, f; Franco-Lara et al. 2013).

Methods of prevention and control

Monitoring pest populations. Direct observation and use of yellow sticky traps are useful methods for monitoring whiteflies and for early detection and the documentation of relative whitefly abundance over time.

Biological control. Despite abundant research on natural enemies and nonchemical control of *B. tabaci* few natural enemies have been identified and used in biological control. Nymphs are affected by entomopathogenic fungi such as *Verticillium lecanii* Zimmerman, *Paecilomyces farinosus* Holmsk, and *P. fomosoroseus* Wize. *V. lecanii* has been developed as a biopesticide (CABI 2017e, f).

Crop management. Maize can be planted as barrier and/or cucurbits as trap crop for controlling adults and to enhance the development of biological control agents (CABI 2017e, f).

Physical control. Yellow attracts whitefly adults. Stationary yellow stick traps are recommended to be installed at planting around fields to capture whitefly adults migrating from other crops. The use of mobile and stationary yellow sticky traps can effectively reduce white fly adult populations in potato (Cisneros and Mujica 1999c; Mujica 1998).

Chemical Control. Avoid use of broad spectrum insecticides in order not to affect natural enemies. The chitin inhibitor, buprofezin, has shown high efficacy (>90%) in controlling immature stages. Soaps (such as potassium soaps) can also be applied on a small scale. Chemical control should be based on using an action threshold of three nymphs per leaf (CABI 2017e, f; Cisneros and Mujica 1999c).

Ladybird Beetles

Henosepilachna vigintioctomaculata (Motschulsky), *Henosepilachna vigintioctopunctata* (F.) (Coleoptera: Chrysomelidae)

Distribution *Henosepilachna vigintioctomaculata* and *H. vigintioctopunctata* are two related and morphologically similar species of phytophagous lady bird beetles that share a common name of a potato ladybird, hadda beetle, or 28-spotted lady bird beetle . *H. vigintioctomaculata* is common in temperate areas of Asia, including China, Japan, Korea, and Russia. The ranges of both species overlap, particularly in China. However, *H. vigintioctopunctata* prefers warmer climates and is widely distributed in Southeast and South Asia, including Pakistan, and has been introduced to Australia, New Zealand, several Pacific islands, and South America (Katakura 1980; Naz et al. 2012; Xu et al. 2013).

Host range Potato ladybirds are polyphagous species feeding on a variety of plant species. However, most severe damage is usually reported for Solanaceae's crops, including potato (Xu et al. 2013).

Symptoms of infestation Both adults and larvae feed on potato leaves, causing their complete defoliation and eventual death in cases of severe infestation. Feeding by larvae usually starts on the lower sides of infested leaves. Lower epidermis and mesophyll get consumed, while higher epidermis and large veins remain relatively intact (Jackson 2016; Xu et al. 2013). Affected leaves eventually dry up and die. Furthermore, feeding damage may facilitate infections by gray mold, *Botrytis cinerea* Pers. ex Fr. (Yao et al. 1992).

Impacts on production losses Potato ladybirds can be serious pests of potato. Yield losses have been estimated to reach 10–15% in normal years, and 20–30% in the years of heavy infestation (Song et al. 2008). In especially severe cases, complete crop destruction is possible (Jackson 2016).

a b

Fig. 28-spotted ladybird beetle, Henosepilachna vigintioctomaculata, adult (**a**) and larva (**b**). (Photo credits: Yulin Gao, Institute of Plant Protection, Chinese Academy of Agricultural Sciences)

Methods of prevention and control

Monitoring pest populations. All life stages of potato ladybirds are rather conspicuous, in large part due to their aposematic coloration. Therefore, they can be relatively easily detected by visual observations. Eggs are usually laid in clusters of 20–30 on lower leaf surfaces, thus requiring that scouted leaves are turned upside-down.

Biological control. The parasitic wasp *Pediobius foveolatus* (Crawford) can parasitize 50–60% of field potato ladybird populations. In some cases, this reduces damage to economically acceptable levels, and no further control is necessary (Puttarudriah and Krishnamurti 1954; Venkatesha 2006). However, such high parasitism rates should not be taken for granted, and a careful monitoring of ladybird populations is required.

Crop management. Removing and destroying crop residue after harvest deprive resident potato ladybirds of food, thus dramatically enhancing their mortality (Jackson 2016; Xu et al. 2013).

Physical control. Manual destruction of potato ladybird beetles is recommended for small potato plots (Xu et al. 2013). Success of this approach is facilitated by the tendency of ladybird adults to aggregate at overwintering sites outside of potato fields. Adult ladybird beetles are strongly phototactic (Zhou et al. 2015). Therefore, light traps can be used to reduce their numbers. However, feasibility of this approach remains to be investigated. Also, light traps attract a wide variety of insects, including beneficial species (Zhou et al. 2015).

Biological insecticides. Some botanical products, such as azadirachtin, have a good efficacy for ladybird management (Ghosh and Chakraborty 2012; Jeyasankar et al. 2014). Also, the entomopathogenic fungi *Beauveria bassiana* and *Metarrhizium anisopliae* (Vishwakarma et al. 2011) and *Bacillus thuringiensis* (Song et al. 2008) are used.

Chemical Control. Potato ladybird beetles can be easily killed by common broad-spectrum insecticides, which remains unfortunately the most common approach to controlling these pests (Xu et al. 2013).

Minor Pests Globally Present

Cutworms

Agrotis ipsilon (Hufnagel, 1776), *Peridroma saucia* (Hübner, 1808), *Agrotis segetum* (Denis & Schiffermüller, 1775) (Lepidoptera: Noctuidae)

Distribution Cutworms are larvae of several noctuid moth species which are cosmopolitan . Most widely distributed are the black cutworm (*Agrotis ipsilon*), the variegated cutworm (*Peridroma saucia*) and the turnip moth (*Agrotis segetum*). Their origin is uncertain but specimens which were used to describe *A. ipsilon* in 1766 were collected from Austria.

Fig. Cutworm larval damage on potato tuber (**a**), and freshly cut potato plant (**b**). Cutworm larva (**c**), pupa (**d**) and adult (**e**) of *Agrotis ipsylon*. (Photo credits: CIP)

Host range Cutworms are polyphagous and have been reported as a pest on nearly all vegetable crops and some cereals. It's a pest in potato, maize, alfalfa, clover (*Trifolium* L.), cotton, rice, sorghum, strawberry, sugar beet, and tobacco. The larvae sometimes prefer to feed on weeds such as bluegrass (*Poa pratensis* L.), curled dock (*Rumex crispus* L.), lambsquarters (*Chenopodium album* L.), yellow rocket (*Barbarea vulgaris* W.T. Aiton); and redroot pigweed (*Amaranthus retroflexus* L.). Considerable damage has also been observed among shrubs and trees such as linden (*Tilia* sp.), wild plum (*Prunus* sp.), crabapple (*Malus* sp.), and lilac (*Syringa vulgaris* L.). Cereals such as oats (*Avena sativa* L.), barley (*Hordeum vulgare* L.), sorghum, maize, and wheat (*Triticum* spp.) are also affected.

Symptoms of infestation The cutworm larvae remain in the soil at the base of the plant during the day. At night, some species cut down the stems of young potato plants, while other species climb the plants and feed on their leaves. Old instar larvae can occasionally tunnel into potato stems disrupting plant growth. Tubers closer to the ground surface may suffer occasional damage. In a single night, as single larva can cut down several potato plants.

Impacts on production losses Cutworms if not detected early and during outbreaks can cause up to 100% crop loss through cutting down of potato plants.

Methods of prevention and control

Monitoring. Use of sex pheromones in white and yellow traps during spring are quite effective in predicting attacks. Baited traps can also be used to monitor larva populations in the field (Sanchez and Vergara 2002).

Cultural control. Deep tillage exposes larvae and pupae to the action of natural enemies and may crash them mechanically. Weed control also aids in reducing cutworm damage by reducing the sites for egg-laying because these insects infest a wide range of host plants. Properly covering tubers during hilling impedes larval access to the tubers. Late harvest results in a greater damage of tubers. Hand picking of the larvae very early in the morning is the most commonly used method for cutworm control among smallholder farmers in Africa. Fallowing and crop rotation with none host plants such as bluegrass, onion, garlic, peppermint (*Mentha piperita* L.) and coriander can also reduce cutworm populations as well as habitat management and farming practices that conserve populations of existing natural enemies. Trap cropping using susceptible host plants such a sunflower or cosmos (*Cosmos* sp.) has also been used in combination with daily killing the trapped cutworm larvae. Flooding of the infested field between crops can also kill the cutworm larvae. Sticky substances such as mollasses and saw dust can also be spread at the base of the plant to trap the cutworm larvae.

Transgenic plants. Several transgenic crops (maize, tobacco, wheat, cotton and maize) have been reported to be resistant to cutworm damage. The genes responsible for resistance against *A. ipsilon* include *Bacillus thuringiensis* cry1Ac gene (Bt), barley trysin inhibitor (Bti-cme) and cowpea trypsin inhibitor (CpTi). Preliminary results of genetically modified potatoes show that feeding larvae on leaves which expresses Cry3Aa affect larvae by being toxic, and also that Bt potatoes curb the growth and reproduction of the adults of *Agrotis ipsilon*.

Biological control. Cutworms are hosts for numerous parasitoid wasps and flies, including species of Braconidae *Cotesia ruficrus* (Haliday), *Snellenius manilae* (Ashmead), Ichneumonidae *Tenichneumon panzer* (Wesmael), Tachinidae *Bonnetia comta* (Fallen), *Euplectrus plathypenae* and Eulophidae, with rates of parasitism as high as 75–80%. Up to 60% of cutworm larvae have been reported to be killed by an entomopathogenic nematode, *Hexamermis arvalis* in central USA. Entomopathogenic nematodes that have been used in the control of *A. ipslon* larvae include *Steinernema glaseri, S. riobrave and Steinernema carpocapsae* (Sanchez and Vergara 2002).

Biopesticides. Extracts from the following plants have demonstrated to be toxic to the larvae of *A. ipslon*: *Nerium oleander* L. leaves, neem leaves and seeds, *Melia azedarach* L. fruits, *Bassia* muricate (L.) Asch., *Lantana* sp., *Parthenium* sp., *Hyptis* sp., *Ipomoea carnea* Jacq., *Tephrosia nubica* (Boiss.) Baker, *Bidens pilosa L.* and *Rumex nepalensis* Spreng. roots (CABI 2018a).

Chemical control. Spotted or localized field infestations by cutworms are typical, calling for focused subsurface soil insecticide applications. The use of toxic baits is recommended because of its specificity against these insects and because it does not markedly affect the biological control agents. Toxic baits prepared as a mix of bran, molasses, water, and insecticide should be applied at the base of plants at dusk (CIP 1996; Larrain et al. 2003).

Armyworms

Spodoptera eridania (Stoll), *Spodoptera frugiperda* (J E Smith), *Mamestra configurata* (Walker), *Spodoptera ornithogalli* Guenée, *Copitarsia decolora* (Guenée), *Feltia* spp., (Lepidoptera: Noctuidae)

Distribution *Spodoptera frugiperda* is distributed in North, Central, and South America, Africa, and India (CABI 2018a). *S. frugiperda* is on the EPPO A1 list of quarantine pests and is intercepted occasionally in Europe on imported plant material (CABI 2018b). *S. eridania* occurs throughout southern USA, Central and South America, and the Caribbean. It is not established in Europe (CABI 2018c). *Mamestra configurata* is restricted to North America (Canada, USA, Mexico) (CABI 2018d).

Host range Army worms feed on a wide range of crops and are important pests especially in cereals. However. they have also been reported to feed on potato in absence of the primary hosts (Strand 2006).

Symptoms of infestation Armyworms are mainly forage feeders and feed during the day, do not make burrows, and may migrate in mass into potato fields from adjacent crops. They also feed on tubers that are exposed on the surface or accessible through cracks in the soil (CIP 1996; Larrain et al. 2003; Sanchez and Vergara 2002).

Impacts on production losses Armyworms can cause up to 100% crop defoliation of potato plants. Some defoliation from armyworms can be tolerated. Keeping defoliation between 10 – 15% will generally prevent yield losses.

Methods of prevention and control It is recommended to make suitable soil preparations that expose the pupa of the insect to environmental conditions and predators. There is biological control based on the use of natural enemies of eggs and larvae, such as Tachinid insects, Inchneumonids, Trichogramatids, the fungus

Fig. *Spodoptera frugiperda* larvae infesting potato leaves (**a**). (Photo credits: CIP)

Zoophthora radicans affecting larvae, among others (Larrain et al. 2003). Spotted or localized field infestations are typical, calling for focused insecticide treatments. Toxic baits prepared as a mix of bran, molasses, water, and insecticide should be applied at the base of plants at dusk (CIP 1996).

Wireworms

***Agriotes lineatus* (L.), *A. vobscurus* (L.), *A. sputator* (L.)., *Athous haemorrhoidalis* (F.), *Conoderus rudis* (Brown), *C. vespertinus* (Fabricius), *Ctenicera pruinina* (Horn), *C. cuprea* (Fabricius), *Melanotus communis* (Gyllenhal) (Coleoptera: Elateridae)**

Distribution Wire worms are the larvae of click beetles. Over 39 species in 21 genera of wireworms have been reported to attack potato. Wireworms are found throughout the world, and species vary greatly among regions. While in the UK, the three most important pest species in potato are *Agriotes lineatus*, *A. vobscurus,* and *A. sputator*, in the US, these are *Limonius canus* LeConte, *Limonius californicus* (Mannerheim), *A. lineatus*, *A. obscurus* L. and *A. sputator.*

Host range Grasses are the main host but they also attack potato, asparagus, carrot (*Daucus carota* subsp. *sativus* (Hoffm.) Schübl. & G. Martens), sugar beet (*Beta vulgaris* L.) and leek (*Allium ampeloprasum* L.)

Symptoms of infestation Wireworms are frequent pests in temperate climates, but less so in warm tropical areas. Thin, lustrous larvae with small thoracic legs live underground and may be up to 25 mm long. The larvae can borrow deeply inside tubers, and spend considerable time over there producing irregularly shaped tunnels in tubers (CIP 1996; Larrain et al. 2003; Sanchez and Vergara 2002).

Impacts on production losses Wireworms may injure potatoes by feeding on the seed piece resulting in weak stands, but most of their damage is caused by tunneling

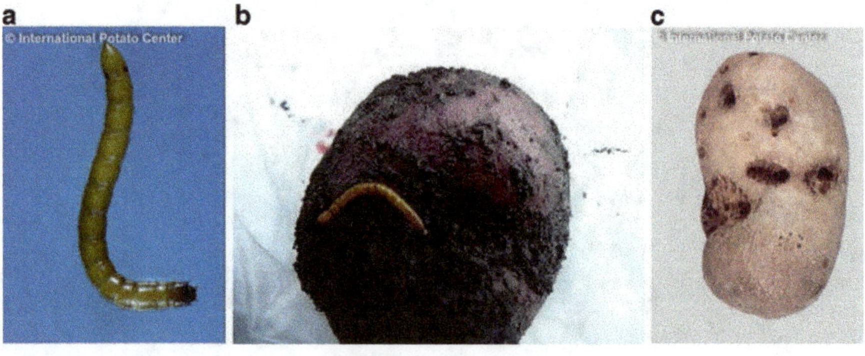

Fig. Wireworm larvae (**a**, **b**) and larval damage on potato tuber (**c**). (Photo credits: CIP)

into tubers, which reduces yield quality. Wireworm tunneling also creates an entry point for certain plant pathogens, eventually leading to tuber rot. In some years and regions of the U.S., up to 45% of the total potato tuber harvest has been downgraded or rejected outright because of wireworm injury, resulting in substantial economic loss (Steele 2011).

Methods of prevention and control
Monitoring. Avoid growing potatoes in wireworm infested fields. Plan and utilize a range of risk assessment methods such as pheromone and bait trapping, as well as soil sampling to confirm the status of each field. In arable rotations, plough-based cultivation may help to reduce wireworm populations (Larrain et al. 2003).

Cultural control. Wireworms feed on the roots of various crops, particularly grasses. Consequently, before planting potatoes in pasture areas, the soil borne wireworm larvae population must be reduced by proper plowing and rotation with other crops that require frequent tilling. Keeping potato fields free from weeds has also been reported to reduce wireworm populations.

Biological control. Currently there are no commercial biological control agents available for controlling wireworms, although some strains of the insect-pathogenic fungus *Metarhizium anisopliae* have shown promising results under experimental conditions (Kabaluk et al. 2005).

Chemical control. Insecticides applied to soil may be required in certain circumstances. To be effective, insecticides for the control of wireworms need to be incorporated into the soil at planting and remain active during the cropping season (Parker and Howard 2001).

Flea Beetles

Epitrix spp. (Coleoptera: Chrysomelidae)

Distribution *Epitrix* spp. is a genus of many flea beetles that are known to feed upon members of the Solanaceae family. *Epitrix tuberis* Gentner (1944) and *E. cucumeris* (Harris 1851) are common pests of potato in North America (Malumphy et al. 2016) . A nonnative pest of potato recently established and causing significant economic damage in Portugal and Spain was identified as *Epitrix papa*

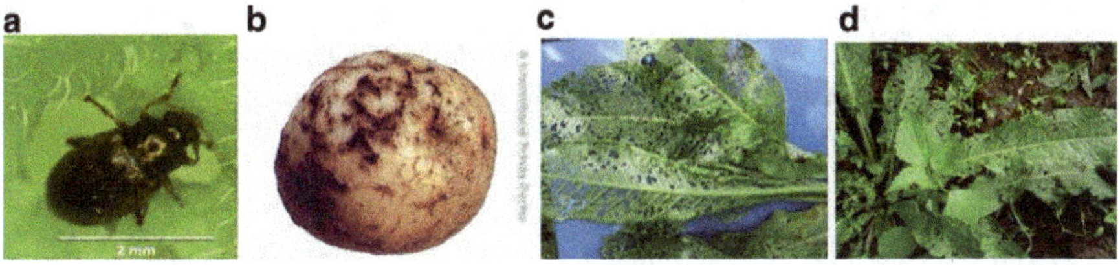

Fig. Adult damage on potato leaves (**a**) and larval damage on potato tuber (**b**). Heavy adult damage and infestation on a weed in potato field in Kabale, Uganda (**c, d**). (Photo credits: CIP)

sp. n. and included in the list of quarantine pests in Europe (Eyre and Giltrap 2013; Malumphy et al. 2016; Sanchez and Vergara 2002). *Epitrix yanazara* Bechyné (1959) was the major economic important species in the central highlands of Peru (Kroschel et al. 2012) which is a region with a high diversity of flea beetles (Furth et al. 2015).

Host range The host range comprises potato, tomato, eggplant, tobacco as well as weeds of the family Solanaceae.

Symptoms of infestation Flea beetles are small beetles and jump easily in the foliage of plants. Their feeding results in characteristic circular holes less than 3 mm in diameter. Larvae that feed on roots, stolons, and tubers also cause damage. They bore the tubers superficially or scratch the skin, thus facilitating penetration of pathogenic fungi (CIP 1996; Larrain et al. 2003; Sanchez and Vergara 2002).

Impacts on production losses The flea beetles are considered one of the most serious pests threatening the entire EPPO region. In North America, they are ubiquitous. In cases of severe infestation, defoliation by adults may cause leaves to dry completely, thus affecting photosynthesis and plant yield. Larval damage makes tubers become unviable for sale, and the destruction of the roots can result in plant death (Eyre and Giltrap 2013; Malumphy et al. 2016; Sanchez and Vergara 2002). In Peru, populations increased severely above economic thresholds if no pesticides were applied (Kroschel et al. 2012).

Methods of prevention and control

Cultural control. Crop rotation. Removal of host weeds and appropriate soil management to ensure vigorous potato plants contribute to reducing flea beetle populations.

Chemical control. Many insecticides that are used to control other insect pests of potato are also effective against flea beetles. Potato plants withstand some foliage damage beyond which insecticides are required (CIP 1996; Larrain et al. 2003; Sanchez and Vergara 2002).

White Grubs

***Phyllophaga* spp., *Anomala* spp., *Popillia japonica* Newman, *Anomala orientalis* Waterhouse, *Holotrichia javana* Brenske, *Holotrichia oblita* Faldermann, *Temnorhynchus coronatus* (Fabricius), *Holotrichia longipennis* Blanchard, *Brahmina coriacea* Hope, *Melolontha melolontha Linnaeus* (Coleoptera: Scarabaeidae)**

White grubs are found in many genera in the family Scarabaeidae.

Distribution Scarab beetles and thus white grubs are found worldwide.

Host range White grubs are polyphagus and feed on over 1000 plant species including field crops such as potato, sweetpotato, maize, asparagus, and soybean;

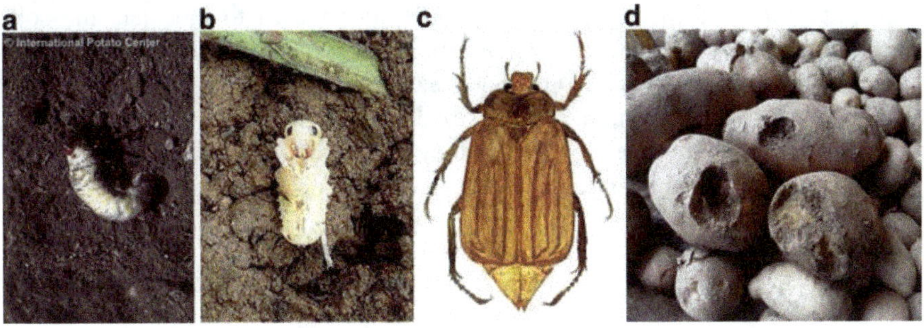

Fig. Typical white grub larva (**a**) and pupa (**b**) of the genus *Phyllophaga*. Adult of *Melolantha melolantha* (**c**) and damage of larvae in potato tubers (**d**). (Photo credits: CIP)

fruits such as apple, peach, grape, as well as trees and ornamentals (Pathania and Chandel 2016; Visser and Stals 2012; Misra and Chandel 2003).

Symptoms of infestation White grubs are the larvae of relatively large beetles. The damage is done mainly by the grubs which remain in the soil. They damage the plant by feeding on roots, underground stem parts, stolons, and tubers. Earlier instars feed on the roots and may cause the plants to dry up. After tubers are developed, the grubs switch to feeding on them, leaving large holes (BioNET-EAFRINET 2011; CIP 1996; Sanchez and Vergara 2002; Visser 2012).

Impacts on production losses Economic damage results mostly from larval damage to potato tubers. Severe damage usually occurs when planting potato in former pasture or grazing fields, which are preferred habitats for white grubs, and on no-till or reduced-till land (CIP 1996; Larrain et al. 2003; Sanchez and Vergara 2002). Yield losses of 40–80% have been reported in India (Pathania and Chandel 2016; Misra 1995).

Methods of prevention and control
Cultural practices. Deep tilling exposes white grubs to adverse environmental conditions such as sunlight and frost, and to predatory vertebrates. Ensuring proper drainage in the field will reduce white grub populations since they prefer moist soil, especially with high content of decaying organic matter. Potato should not be planted directly following pasture, grass, or several successive years of cereals and/or corn. Crop rotation is an effective control method when potato is rotated with resistant or less susceptible crops, such as deep-rooted legumes (e.g., alfalfa, cowpea, and pigeon peas) and Allium crops (onions, garlic). Strips of African marigold (*Tagetes* sp.), sunflower (*Helianthus annuus* L.), and castor can be used as trap crops. Allium crops also repel adult beetles from laying eggs (Larrain et al. 2003; BioNET-EAFRINET 2011).

Biological control. Natural enemies that control white grubs include parasitic wasps in the genera *Tiphia, Myzinum* (Hymenoptera: Tiphiidae), and *Pelecinus polyturator* Drury (Hymenoptera: Pelecinidae), and the fly, *Pyrgota undata* Wiedemann (Diptera: Pyrgotidae). Spores of the pathogens *Bacillus popilliae*,

B. lentimorbus, and *Metarhizium anisopliae* can be used to inoculate the soil. Nematodes species such as *Steinernema* sp. and *Heterrorhabditis* sp. can also be effectively used against white grubs (Sanchez and Vergara 2002).

Chemical control. This pest is not easily controlled with insecticides. Chemical treatment on grubs must be done when they are young, as older larvae are more robust and move to a greater depth as they develop (Gupta and Gavkare 2014).

Mites

Tetranychus urticae C. L. Koch, *T. evansi* Baker & Prichard, *Polyphagotarsonemus latus* (Banks) (Arachnidae: Trombidiformis, Tarsonemidae)

Distribution The origin of the tomato or tobacco red-spider mite *Tetranychus evansi* is Brazil. The pest was accidentally introduced into Southern Africa during the 1980s. The two spotted- or red-spider mite *Tetranychus urticae* and the broad mite *Polyphagotarsonemus latus* are widely distributed in both tropical and subtropical areas of the world (CABI 2017g, h).

Host range Mites are polyphagous. Major hosts are potato, tomato, tobacco, pepper, eggplant, pumpkins and squashes (*Cucurbita pepo* L.), cucumber, watermelon, celery, maize, beans, strawberry (*Fragaria* × *ananassa*), cotton, citrus, papaya (*Carica papaya* L.) (Hazzard 2013; Landis and Davis 1947; Goftishu et al. 2016).

Symptoms of infestation *T. urticae* and related mites are generally known as red spiders, although red is not always their characteristic color. Mites are extremely small, almost microscopic, and feed on the cellular matter of leaves. Chlorotic spots caused by mites give leaves a tan coloring, whereas high infestation will cause leaf and plant wilting. *P. latus* attacks sprouts and tender leaves, deforming them. Damage to growing plants is particularly severe. The white mite colonies are located on the underside of the young leaves and on the buds. The infested plants take a dark

Fig. *Tetranychus urticae* feeding on a potato leaf (**a**), and symptoms of white mite feeding causing deformation of buds and tender leaves (**b**) and toasted appearance on the underside of a potato leaf (**c**). (Photo credits: CIP)

to moderate green coloration and develop a somewhat leathery appearance. Heavy infestation may result in plant death (CIP 1996; Larrain et al. 2003; Sanchez and Vergara 2002; Landis and Davis 1947).

Impacts on production losses Plants with severe damage of *P. latus* do not form tubers and remain very small. When infestations occur in an intense and violent way they can cause the destruction of entire crop field. In the same way, attack by hundreds or thousands of red spider mites can cause thousands of lesions, thus can significantly reduce the photosynthetic capability of plants (CIP 1996; Larrain et al. 2003; Sanchez and Vergara 2002).

Methods of prevention and control
Monitoring. Mite infestations normally begin at boundaries of the potato fields especially those neighboring fields of preferred host plants such as maize, alfalfa, and mint. Stressed potato plants tend to be more susceptible. Mites can build high populations in a very short time during hot (above 30 °C) dry seasons. Constant visual monitoring and early detection is key. Check on the underside of leaves that show symptoms of nutrient deficiency (CABI 2017g, h).

Cultural practices. Sprinkler irrigation helps to limit mite damage by increasing the humidity (above 60%) on plant leaves, thus making conditions less favorable for mites. Proper field sanitation is done through removal of infested plants and volunteer crops (CIP 1996; CABI 2017g, h).

Natural control. Often mites are kept under control by their natural enemies, including predatory thrips, lacewings, gall midges (*Feltiella acarisuga* Vallot) and ladybird beetles (*Stethorus punctillum* Weise) and predatory mites (*Phytoseiulus persimilis* Athias-Henriot, *Neoseiulus californicus* (McGregor), *Amblyseius andersonii* Berlese, *Galendromus occidentalis* Nesbitt, *Mesoseiulus longipes*, and *Hypoaspis miles* Berlese (Pundt 2007). Mites become primary pest when natural enemies are destroyed by insecticide applications to control other insect pests. So, avoid use of nonselective pesticides which reduce predatory mite populations.

Biopesticides. Plant extracts such as azadirachtin (neem oil), or from cotton seed, soybean, clover, garlic, rosemary (*Rosemarinus officinalis* L.), and pyrethrum have shown some efficacy against red spider mites (CABI 2017g, h).

Chemical control. When control measures are required, usually a single application of an acaricide is sufficient especially when populations are still low (<five mites per plant). In some cases, focal treatments are recommended using either soap, sulphur, Bifenazate, abamectin, spirotetramat, and spiromesifen (Hazzard 2008).

Thrips

Frankliniella occidentalis (Pergande), *Thrips tabaci* Lindeman (Thysanoptera, Thripidae)

Distribution The western flower thrips *Frankliniella occidentalis* originated in western North America and has since become a major pest on many crops across the

Fig. Pale or brown nymphs and darker adults of *Frankliniella occidentalis* on the underside of potato leaves (**a**), and symptoms of thrips feeding causes silvering on underside of a potato leaf (**b**). (Photo credits: CIP)

US and around the world. The onion thrip *Thrips tabaci* is thought to have originated in the Mediterranean region but is now found on all continents except Antarctica.

Host range *F. occidentalis* feeds on more than 200 crops in over 50 genera and this includes vegetables, fruit trees, and ornamentals.

Symptoms of infestation Thrips can cause major damage mainly during dry climatic conditions. Because of heavy feeding, potato leaves develop a silver or chlorotic dotting of the tissues and become deformed. The silver color is due to the emptying of the cells (CIP 1996; CABI 2017i, j; Learmonth 2017).

Impacts on production losses Thrips are minor pest of potato. Some species are vectors for tomato spotted wilt virus (TSWV), with western flower thrips (*F. occidentalis*) as the most important virus vector. The virus reduces crop yield and tuber quality (CABI 2017i, j).

Methods of prevention and control
Cultural practices. Thrips feed on a wide range of plants, including weeds from which they can invade potato crops. Therefore, weed control is a good management strategy for this pest. Susceptible crops (onion, tomato, corn, artichoke, and pumpkin) should not be planted next to a potato crop. Any plants showing tomato spotted wilt virus symptoms should be removed (CIP 1996; CABI i, j; Learmonth 2017).

Behavioral control. Yellow or blue (especially for *F. occidentalis*) sticky traps can be used to monitor the presence of thrips.

Biological control. Various species of the worldwide anthocorid genus *Orius* are used in biological control against thrips, and these bugs are important as predators in many natural populations. *Amblyseius swirskii* Athias-Henriot and *Neoseiulus cucumeris* Oudemans are two of the most widely used predatory mites in the biological control of *F. occidentalis*. Fungal pathogens and nematodes, such as *Beauveria bassiana* and *Steinernema feltiae*, are also being used commercially. Parasitoids include *Ceranisus menes* Walker for the western flower thrip (CABI 2017i, j).

Chemical control. Low to medium toxicity pesticides used against thrips include Spinosad, Spinetoram, and Abamectin (Bentley and Rice 2015).

Tomato Tuber Moth

Tuta absoluta (Meyrick) (Lepidoptera: Gelechiidae)

Distribution The tomato leaf miner, *Tuta absoluta*, was first identified in Peru by Meyrick in the year 1917 from samples collected in Huancayo (3200 m above sea level) . Since then, it has been recorded in all South American countries.
The pest is predominantly found in areas below 1000 m asl. It was first introduced into Europe in Spain in 2006, from where it is believed to have spread to other European countries. Its range has expanded considerably to Asian and African countries in recent years (CABI 2017k).

Host range Hosts of *T. absoluta* are the cultivated Solanaceae tomato, potato, egg-plant, sweet cucumber (*Solanum muricatum* L.), and tobacco. Other cultivated host plants are slender amaranth (*Amaranthus viridis* L.), groundcherry, and the Johnson grass (*Sorghum halepense* (L.) Pers.). Noncultivated Solanaceae hosts include a wide range of weeds such as *Lycopersicon hirsutum* Dunal, *Lycopersicum puberulum* Ph, *Solanum americanum* Mill., *S. bonariense* L., *S. elaeagnifolium* L., *S. hirtum*, *S. lyratum*, *S. nigrum* L., *S. puberulum*, *S. sisymbriifolium* Lam., *S. saponaceum*, *Datura stramonium* L. (jimsonweed), *D. ferox* L., *Nicotiana glauca* Graham, *Lycium chilense* (Bertero), *Lycopersicon pennellii* var. *puberulum* (Correll), *Solanum dubium* Fresen, and *Solanum puberulum* Phil.

Symptoms of infestation On potato, *T. absoluta* attacks aerial parts and occasionally tubers. Foliar damage is a result of mine-formation within the mesophyll by the feeding larvae. The newly emerged larvae are quite mobile and begin their feeding by building galleries on the leaflets. Depending on the place of oviposition, a larva may penetrate leaf mesophyll, leaf petiole, and young shoots. The mines expand as

Fig. Adult of *Tuta absoluta* (**a**), larvae infestation on potato leaf (**b**), and tuber (**c**). (Photo credits: CIP)

the larva develops, causing deformation or rot of the affected leaves. The larvae can migrate to another part of the plant, especially when they reach their maximum development or when the affected foliage withers and dries (CIP 1996; Notz 1992; EPPO 2005b).

Impacts on production losses *T. absoluta* is a major pest of tomato (*Lycopersicum esculentum*), but it also attacks potato. Larval feeding causes desiccation of damaged leaves and shoots, resulting in low tuber production. Larval damage is generally accentuated in young plants in the tuberization phase. Possibilities of *T. absoluta* infestation of potato tubers are very low and restricted to very high field infestation rates during the last stage of the crop (CIP 1996; CABI 2017k; OIRSA 2015b; SENASICA 2016).

Methods of prevention and control
Monitoring with pheromone traps. Pheromone traps can be used both for early detection and for monitoring the flight activity of the pest. They can also help to reduce the population when used for mass trapping, in which case a higher number of traps need to be deployed. For monitoring purposes, one trap should be placed per hectare (Niederwieser 2017; OIRSA 2015b).

Cultural practices. Good agricultural practices for the control of *T. absoluta* include crop rotation with non-solanaceous crops (preferably cruciferous crops), ploughing, adequate irrigation and fertilization, removal of infested plants, and complete removal of post-harvest plant debris.

Biological control. Nesidiocoris tenuis Reuter is an efficient predator of eggs and small larvae of the tomato tuber moth. Egg parasitoids of the family Trichogrammatidae are the natural enemies most used worldwide in biological control programs against Lepidoptera, through inoculative and inundative releases. Releases of *Trichogramma pretiosum* Riley reduced the population of *T. absoluta* to a maximum of 49%. Therefore, combinations with the application of a bioinsecticide based on *Bacillus thuringiensis* subsp. *kurstaki* have been recommended (Niederwieser 2017).

Botanicals/Biopesticides. Neem oil (Azadirachtin) acts as a contact and systemic insecticide and has been shown to be effective against low infestations of *T. absoluta* and for controlling first and second instar larva. The use of *Bacillus thuringensis* is recommended at low-medium infestation levels in combination with Azadirachtin (SENASICA 2016).

Chemical control. Several insecticides have been effectively used against *T. absoluta* in potato; however, this pest is known to develop resistance against effective insecticides.

Potato Leafhoppers

Empoasca fabae (Harris 1841) (Hemiptera: Cicadellidae)

Distribution The leafhopper *Empoasca fabae* is widely distributed.

Fig. *Empoasca fabae* adult (**a**); and potato leave damage (**b**). (Courtesy by CIP)

Host range *E. fabae* is a polyphagous pest. Its main hosts are potato, alfalfa, eggplant, bean, celery, cucumber, cucurbits, groundnut (*Arachis hypogaea* L.), melon, rhubarb (*Rheum rhabarbarum* L.), strawberry, sweet potato, and tomato.

Symptoms of infestation The potato leafhopper *E. fabae* is a sucking insect that extracts sap directly from the vascular system of the leaflets, petioles and sometimes the stems. The attack on potato crops is sporadic and is favored by humid conditions. Insects live in the abaxial (lower) part of the leaflets. They inject toxic saliva while feeding, causing leaf necrosis and interfering with plant growth. Under severe attack, the attacked plants die prematurely. Leafhopper can transmit some viruses, although their occurrence is rare (CIP 1996; Cook et al. 2004; Larrain et al. 2003).

Impacts on production losses The complex of symptoms caused by the leafhopper leads to reduced growth and reproduction of plants. Depending on the stage of plant development, in heavily infested fields up to 75% the yield can be lost (Backus et al. 2005; Cook et al. 2004; Medeiros et al. 2004). *E. fabae* has become a major pest with the introduction of modern synthetic insecticides. In Minnesota, USA, annual losses to potato were estimated at $US 7 million (Noetzel et al. 1985). The relationship between yield loss and leafhopper numbers have been determined to be directly linear (Radcliffe and Johnson 1994). In south western Ontario, Canada, average losses of up to 85% for potatoes have been recorded (Tolman et al. 1986). *E. fabae* is also a serious pest of potato in some parts of India with severe hopper burn especially of early planted crops (Verma et al. 1994).

Methods of prevention and control

Most of the potato leafhopper management in potato and other crops is based on monitoring, cultural control, and the use of insecticides.

Monitoring. The presence of the leafhopper can be determined rapidly using entomological nets or yellow sticky traps located at the edges of the crop. Nymphs are best monitored by examining the leaves, especially the underside. Monitoring middle part of a plant gives a more precise estimate of the population of leafhopper nymphs than monitoring basal and apical parts. The economic threshold in potato is 10 or more nymphs per 100 leaves, or 10% of the leaves infested with nymphs (CIP 1996; Cook et al. 2004).

Cultural practices. Infestation of potato fields may be prevented by avoiding proximity to crops such as beans that host high leafhopper populations. Also, elimination of crop residues and appropriate irrigation help reduce populations of this pest (Larrain et al. 2003).

Natural control. Although the leafhopper has several natural enemies, such as predators and parasites, they play a very minor role in potato leafhopper control (CIP 1996; Cook et al. 2004; Larrain et al. 2003).

Plant resistance. Resistant or tolerant varieties should be considered in affected regions, but if leafhopper populations increase, systemic insecticides may become necessary (Backus et al. 2005).

Chemical control. Applications must be made only if populations reach economic thresholds.

Leaf Beetles

***Diabrotica viridula* Fabricius, *D. speciosa* Erichson (Coleoptera: Chrysomelidae)**

Distribution Central America (Costa Rica and Panama) and South America (Argentina, Bolivia, Brasil, Columbia, Ecuador, French Guyana, Paraguay, Peru, Uruguay, and Venezuela.

Host range South American Diabrotica species are presented in at least 116 species in 24 families. *Diabrotica speciosa* larvae developed well on maize, peanut (*Arachis hypogaea* L.), and soybean roots, and not so well on pumpkin (*Cucurbita maxima* Duchesne and *Cucurbita andreana* Naudin), beans (*Phaseolus* spp.), and potato roots. *Diabrotica viridula*, preferred maize as adult and larval food, and for oviposition (Cabrera 2003).

Symptoms of infestation Feeding on leaflets by adults interferes with photosynthesis and, therefore, reduces production of tubers. Larvae damage stolons and form galleries in tubers (Cabrera 2003; Lara et al. 2004).

Fig. Adult of *Diabrotica speciosa*. (Photo credits: CIP)

Impacts on production losses In some areas, the most important economic damage to potato is caused by the subterranean larvae gnawing the surface of tubers, which lose quality and become susceptible to soil pathogens (CIP 1996). Eggs and larvae do not develop under dry conditions, so damage is most severe during wet seasons.

Methods of prevention and control
Cultural practices. Deep tillage leads to exposing of larvae and pupae to the action of predators and adverse environmental factors. Weeds that serve as alternating hosts should be eliminated (CIP 1996; Sanchez and Vergara 2002).

Chemical control. In cases of heavy infestation, especially when potato plants are small, insecticide applications are recommended against adult beetles (CIP 1996; Sanchez and Vergara 2002).

Concluding Remarks

Since potato is a major crop for humankind, it has a global distribution and it is attacked by a myriad pests which can substantially reduce its productivity and its quality. The increasing awareness about the nutritional, agronomic, and cash creating advantages potato provides is likely to further increase its status as a global crop, particularly in developing subtropical and tropical countries. The development, adaptation and use of integrated pest management will be an important area of future research crucial for a sustainable and more resilient and economic profitable potato production in all potato growing regions worldwide. Emphasis should be given to develop and use biological approaches in pest management. This will reduce the dependence on insecticides as well as will reduce the risk that insect populations develop resistance against insecticides.

Previous work has clearly shown that many insect pests of potato will respond to climate change by expanding their geographical range of distribution and increasing population densities will lead to greater crop and post-harvest losses, particularly in subtropical and tropical regions. Modeling of the response of potato pest populations to global warming will help to predict potential changes in pest distribution and severity in order to support potato growers in the adaptation of their pest management strategies.

References

Alcázar J, Cisneros F (1999) Taxonomy and bionomics of the Andean potato weevil complex: *Premnotrypes* spp. and related genera. In: Arthur C, Ferguson P, Smith B (eds) Impact on a changing world: program report 1997-98. International Potato Center, Lima, pp 141–151. http://www.cipotato.org/market/PgmRprts/pr97-98/17weevil. Accessed 11 Jun 2018

Alcázar J, Kroschel J (2008) Andean potato weevil. In: Wale SJ, Platt HW, Cattlin ND (eds) Diseases, pests and disorders of potatoes: a color handbook. Academic Press, Boston, pp 129–131

Alyokhin A (2009) Colorado potato beetle management on potatoes: current challenges and future prospects. In: Tennant P, Benkeblia N (eds) Potato II. Fruit, vegetable and cereal science and biotechnology, vol 3 (Special Issue 1), pp 10–19

Alyokhin A, Chen YH (2017) Adaptation to toxic hosts as a factor in the evolution of insecticide resistance. Curr Opinion Insect Sci 21(1):33–38

Alyokhin A, Chen YH, Udalov M, Benkovskaya G, Lindström L (2013) Evolutionary considerations in potato pest management. In: Giordanengo P, Vincent C, Alyokhin A (eds) Insect pests of potato: global perspectives on biology and management. Academic Press, Oxford, pp 543–571

Alyokhin A, Mota-Sanchez D, Baker M, Snyder WE, Menasha S, Whalon M, Dively G, Moarsi WF (2015) Red Queen on a potato field: IPM vs. chemical dependency in Colorado potato beetle control. Pest Manage Sci 71(3):343–356

Anderson PK et al (2005) Whitefly and whitefly-borne viruses in the tropics: building a knowledge base for global action. International Center for Tropical Agriculture (CIAT), Cali, Colombia, p 351

Arthropod Pesticide Resistance Database (2018). https://www.pesticideresistance.org. Accessed 3 Jul 2018

Backus EA, Serrano MS, Ranger CM (2005) Mechanisms of hopperburn: an overview of insect taxonomy, behavior, and physiology. Annu Rev Entomol 50:125–151

Baker WA, Bradley WG, Clark CA (1949) Biological control of the European corn borer in the United States. Tech Bull U S Dep Agri 983:1–185

Bentley WJ, Rice RE (2015) UC IPM pest management guidelines: Nectarine UC IPM Program, Kearney Agricultural Center, Parlier, USA. http://ipm.ucanr.edu/PMG/r540300411.html. Accessed 2 Nov 2018

BioNET-EAFRINET (2011) White grubs. Keys and fact sheets. http://keys.lucidcentral.org

Blackman RL, Eastop VF (2000) Aphids on the world's crops: an identification and information guide, 2nd edn. Wiley, New York

Butler CD, Trumble JT (2012) The potato psyllid, *Bactericera cockerelli* (Sulc) (*Hemiptera: Triozidae*): life history, relationship to plant diseases, and management strategies. Terr Arthropod Rev 5:87–111. https://doi.org/10.1163/187498312X634266

CABI (2012) *Liriomyza huidobrensis* (Blanchard). Invasive species compendium: datasheets, maps, images, abstracts and full text on invasive species of the world. http://www.cabi.org/isc/datasheet/30956. Accessed 2 Feb 2018

CABI (2016) *Bactericera cockerelli* (tomato/potato psyllid). http://www.cabi.org/isc/datasheet/45643. Accessed 20 Feb 2018

CABI (2017a) *Leptinotarsa decemlineata* (Colorado potato beetle) data sheet. In: Invasive species compendium. CAB International, Wallingford, UK. http://www.cabi.org/isc/datasheet/30380. Accessed 25 Feb 2018

CABI (2017b) *Ostrinia nubilalis* (European maize borer). In: Invasive species compendium. CAB International, Wallingford, UK. http://eol.org/pages/464566/details. Accessed 4 Feb 2018

CABI (2017c) *Myzus persicae* (green peach aphid). In: Invasive species compendium. CAB International, Wallingford, UK. http://www.cabi.org/isc/datasheet/35642. Accessed 20 Feb 2018

CABI (2017d) *Macrosiphum euphorbiae* (potato aphid). In: Invasive species compendium. CAB International, Wallingford, UK. http://www.cabi.org/isc/datasheet/32154. Accessed 5 Feb 2018

CABI (2017e) *Bemisia tabaci* [original text by Andrew Cuthbertson]. In: Invasive species compendium. CAB International, Wallingford, UK. http://www.cabi.org/isc/datasheet/8927. Accessed 6 Feb 2018

CABI (2017f) *Trialeurodes vaporariorum*. In: Invasive species compendium. CAB International, Wallingford, UK. http://www.cabi.org/isc/datasheet/54660. Accessed 7 Feb 2018

CABI (2017g) *Tetranychus urticae* (two-spotted spider mite) data sheet. In: Invasive species compendium. CAB International, Wallingford, UK. http://www.cabi.org/isc/datasheet/53366. Accessed 8 Feb 2018

CABI (2017h) *Polyphagotarsonemus latus* data sheet. In: Invasive species compendium. CAB International, Wallingford, UK. http://www.cabi.org/isc/datasheet/26876. Accessed 9 Feb 2018

CABI (2017i) *Frankliniella occidentalis* (western flower thrips). In: Invasive species compendium. CAB International, Wallingford, UK. https://www.cabi.org/ISC/datasheet/24426. Accessed 5 Feb 2018

CABI (2017j) *Trips tabaci* (onion thrips). In: Invasive species compendium. CAB International, Wallingford, UK. http://www.cabi.org/isc/datasheet/53746. Accessed 11 Feb 2018

CABI (2017k) *Tuta absoluta* datasheet. In: Invasive species compendium. CAB International, Wallingford, UK. http://www.cabi.org/isc/datasheet/49260. Accessed 12 Feb 2018

CABI (2018a) Agrotis ipsilon (black cutworm) datasheet. In: Invasive species compendium. Wallingford, UK: CAB International. https://www.cabi.org/isc/datasheet/3801. Accessed 10 Dec 2018

CABI (2018b) *Spodoptera frugiperda* (fall armyworm) datasheet. In: Invasive species compendium. CAB International, Wallingford, UK. https://www.cabi.org/isc/datasheet/29810. Accessed 10 Dec 2018

CABI (2018c) *Spodoptera eridania* (southern armyworm). In: Invasive species compendium. CAB International, Wallingford, UK. https://www.cabi.org/ISC/datasheet/44518. Accessed 10 Dec 2018

CABI (2018d) *Mamestra configurata* (bertha armyworm). In: Invasive species compendium. CAB International, Wallingford, UK. https://www.cabi.org/isc/datasheet/8492. Accessed 10 Dec 2018

Cabrera G (2003) Host range and reproductive traits of *Diabrotica speciosa* (Germar) and *Diabrotica viridula* (F.) (*Coleoptera: Chrysomelidae*), two species of South American pest rootworms, with notes on other species of *Diabroticina*. Environ Entomol 32(2):276–285. https://doi.org/10.1603/0046-225X-32.2.276

Canedo, V, Carhuapoma P, López E, Kroschel J (2016a) *Copidosoma koehleri* (Blanchard 1940). In: Kroschel J, Mujica N, Carhuapoma P, Sporleder M (eds) Pest distribution and risk atlas for Africa. Potential global and regional distribution and abundance of agricultural and horticultural pests and associated biocontrol agents under current and future climates. International Potato Center (CIP), Lima (Peru), pp. 208–219. https://doi.org/10.4160/9789290604761-16. ISBN 978-92-9060-476-1

Canedo V, Carhuapoma P, Dávila W, Kroschel J (2016b) *Apanteles subandinus* (Blanchard 1947). In: Kroschel J, Mujica N, Carhuapoma P, Sporleder M (eds) Pest distribution and risk atlas for Africa. Potential global and regional distribution and abundance of agricultural and horticultural pests and associated biocontrol agents under current and future climates. International Potato Center (CIP), Lima (Peru), pp 220–231. https://doi.org/10.4160/9789290604761-17

Canedo V, Carhuapoma P, Bartra C, Kroschel J (2016c) *Orgilus lepidus* (Muesebeck 1967). In: Kroschel J, Mujica N, Carhuapoma P, Sporleder M (eds) Pest distribution and risk atlas for Africa. Potential global and regional distribution and abundance of agricultural and horticultural pests and associated biocontrol agents under current and future climates. International Potato Center (CIP), Lima (Peru), pp 232–244. https://doi.org/10.4160/9789290604761-18

Capinera JL (2001) Handbook of vegetable pests. Academic Press, San Diego, p 729

Castillo JR (2010) *Prodiplosis longifila* in Peru. https://conference.ifas.ufl.edu/TSTAR/presentations/Tuesday/pm/4%2020pm%20J%20Castillo.pdf. Accessed 20 Feb 2018

CIP (1996) Major potato diseases, insects, and nematodes. International Potato Center, Lima, Peru, p 61

Cisneros F, Mujica N (1999a) Impacto del cambio climático en la agricultura: Efectos del fenómeno El Nino en los cultivos de la costa central. Consejo Nacional del Ambiente (CONAM), Lima (Peru), pp 115–133

Cisneros F, Mujica N (1999b) The leafminer fly in potato: plant reaction and natural enemies as natural mortality factors. International Potato Center, Lima, Peru, pp 129–140

Cisneros F, Mujica N (1999c) Biological and selective control of the sweet potato whitefly, *Bemisia tabaci* (*Gennadius*) (Hom.: *Aleyrodidae*). International Potato Center, Lima, Peru. Program Report 1997-98, pp 255–264

Cook KA, Ratcliffe ST, Gray ME, Steffey KL (2004) Potato leafhopper (*Empoasca fabae* (Harris)). Insect Fact Sheet, University of Illinois, Department of Crop Sciences. https://ipm.illinois.edu/fieldcrops/insects/potato_leafhopper.pdf

Dillon ME, Wang G, Huey RB (2010) Global metabolic impacts of recent climate warming. Nature 467:704–706

EPPO (2005a) *Tecia solanivora*. EPPO Bull 35:399–401

EPPO (2005b) *Tuta absoluta*. Data sheets on quarantine pests. Bull OEPP/EPPO Bull 35:434–435

EPPO (2014) PQR database. European and Mediterranean Plant Protection Organization, Paris, France. http://www.eppo.int/DATABASES/pqr/pqr.htm

EPPO (2015) *Prodiplosis longifila* (*Diptera: Cecidomyiidae*). In: EPPO technical document No. 1068, EPPO study on pest risks associated with the import of tomato fruit. EPPO Paris, pp 133–134. https://www.eppo.int/QUARANTINE/DT_1068_Tomato_study_MAIN_TEXT_and_ANNEXES_2015-01-26.pdf

European Parliament and the Council of the European Union (2009) Directive2009/128/EC of the European Parliament and of the Council of 21 October 2009 establishing a framework for Community action to achieve the sustainable use of pesticides. Official Journal of the European Union, L309/71. First published 2009. http://eur-lex.europa.eu/legal-content/EN/ALL/?uri=celex%3A32009L0128

Eyre D, Giltrap N (2013) *Epitrix* flea beetles: new threats to potato production in Europe. Pest Manage Sci 69(1):3–6

FAO (2018). http://www.fao.org/agriculture/crops/core-themes/theme/pests/ipm

Ferro DN (2000) Success and failure of Bt products: Colorado potato beetle—a case study. In: Kennedy GG, Sutton TB (eds) Emerging technologies for integrated pest management: concepts, research, and implementation. APS Press, St. Paul, pp 177–189

Forgash AG (1985) Insecticide resistance in the Colorado potato beetle. In: Ferro DN, Voss RH (eds) Proceedings, Symposium on the Colorado potato beetle. XVIIth International Congress of Entomology. Res. Bull. 704, Mass Agric Exp Stn Circ 347, pp 33–52

Franco-Lara L, Rodríguez D, Guzmán-Barney M (2013) Prevalence of potato yellow vein virus (PYVV) in *Solanum tuberosum* group *Phureja* fields in three states of Colombia. Am J Potato Res 90(4):324–330

Furth DG, Savini V, Chaboo CS (2015) Beetles (Coleoptera) of Peru: a survey of the families. Chrysomelidae: Alticinae (flea beetles). J Kans Entomol Soc 88(3):368–374

Gagné RJ (1986) Revision of *Prodiplosis* (*Diptera: Cecidomyiidae*) with descriptions of three new species. Ann Entomol Soc Am 79(1):235–245

Gamarra H, Mujica N, Carhuapoma P, Kreuze J, Kroschel J (2016a) Sweetpotato white fly, *Bemisia tabaci* (Gennadius 1989) (Biotype B). In: Kroschel J, Mujica N, Carhuapoma P, Sporleder M (eds) Pest distribution and risk atlas for Africa—potential global and regional distribution and abundance of agricultural and horticultural pests and associated biocontrol agents under current and future climates. International Potato Center (CIP), Lima, Peru, pp 85–99. https://doi.org/10.4160/9789290604761 (online publication: http://cipotato.org/riskatlasforafrica/)

Gamarra H, Carhuapoma P, Mujica N, Kreuze J, Kroschel J (2016b) Greenhouse whitefly, *Trialeurodes vaporariorum* (Westwood 1956). In: Kroschel J, Mujica N, Carhuapoma P, Sporleder M (eds) Pest distribution and risk atlas for Africa—potential global and regional distribution and abundance of agricultural and horticultural pests and associated biocontrol agents under current and future climates. International Potato Center (CIP), Lima, Peru, pp 154–168. https://doi.org/10.4160/9789290604761 (online publication: http://cipotato.org/riskatlasforafrica/)

Garzón TJA (2002) Asociación de *Paratrioza cockerelli* Sulc. con enfermedades en papa (*Solanum tuberosum*) y tomate (*Lycopersicum lycopersicum* Mil. Ex. Fawnl) en México. In: Memoria del Taller sobre 11 *Paratrioza cockerelli* (Sulc.) como plaga y vector de fitoplasmas en hortalizas. Culiacán, Sinaloa, México, pp 79–87

Ghosh SK, Chakraborty G (2012) Integrated field management of *Henosepilachna vigintiocto-punctata* (Fabr.) on potato using botanical and microbial pesticides. J Biopest 5(Suppl):151–154

Goftishu M, Dejene M, Kassaye A, Belay T (2016) Red Spider Mite, *Tetranychus urticae* Koch (Arachnida: Acari-Tetranychidae): a threatening pest to potato (*Solanum tuberosum*). Pest Manage J Ethiopia 19:53–59

Gonzales-Bustamante L (1996) *Prodiplosis longifila* infestando vainas del pallar en Lima. Rev Peru Entomol 39:118

Gupta S, Gavkare O (2014) White grub, *Brahmina coriacea*, a potential threat to potato. J Ind Pollut Control 30(2):357–359

Hazzard R (2008) New England vegetable management guide. New England extension systems, University of Massachusetts, Amherst, MA. https://ag.umass.edu/vegetable/fact-sheets/two-spotted-spider-mite. Accessed 2 Nov 2018

Hazzard R (2013) New England vegetable management guide. New England extension systems, University of Massachusetts, Amherst, MA. https://ag.umass.edu/vegetable/fact-sheets/two-spotted-spider-mite. Accessed 23 Sep 2019

Henne DC, Paetzold L, Workneh F, Rush CM (2010) Evaluation of potato psyllid cold tolerance, overwintering survival, sticky trap sampling, and effects of liberibacter on potato psyllid alternate host plants. In: Workneh F, Rush CM (eds), Proceedings of the 10th Annual Zebra Chip Reporting Session, pp. 149–153. Dallas, TX (November 7–10, 2010)

Hernandez LM (2014) Caracterización del daño y distribución geográfica de Cecidomyiidae (Diptera) y sus parasitoides asociados a Solanáceas y limón Tahití en Colombia. Tesis de Maestría en Ciencias Agrarias. Línea Protección de Cultivos. Universidad Nacional de Colombia. Palmira, 93 p

Hernandez LM, Guzman YC, Martínez-Arias A, Manzano MR, Selvaraj JJ (2015) The bud midge *Prodiplosis longifila*: damage characteristics, potential distribution and presence on a new crop host in Colombia. SpringerPlus 4:205. https://doi.org/10.1186/s40064-015-0987-6

Horn P, Page J (2008) Integrated pest management dealing with potato tuber moth and all other pests in Australian potato crops. In: Kroschel J, Lacey L (eds) Integrated pest management for the potato tuber moth, *Phthorimaea operculella* (Zeller)—a potato pest of global importance. Tropical Agriculture 20, Advances in Crop Research 10. Margraf Publishers, Weikersheim, Germany, pp 111–117

Jackson G (2016) Potato ladybird beetle (255). Pacific pests and pathogens fact sheet. PestNet, Queensland, Australia

Jacques RL, Fasulo TR (2015) Colorado potato beetle *Leptinotarsa decemlineata*. DPI Entomology Circular 271. Florida Department of Agriculture and Consumer Services, Division of Plant Industry and University of Florida. http://entnemdept.ufl.edu/creatures/veg/leaf/potato_beetles.htm

Jeger M, Bragard C, Caffier D, Candresse T, Chatzivassiliou E, Dehnen-Schmutz K, Gilioli G, Gregoire J-C, Jaques Miret JA, Navajas Navarro M, Niere B, Parnell S, Potting R, Rafoss T, Rossi V, Urek G, Van Bruggen A, Van der Werf W, West J, Winter S, Gardi C, Bergeretti F, MacLeod A (2018) Scientific opinion on the pest categorisation of *Tecia solanivora*. EFSA J 16(1):5102,. pp 25. https://doi.org/10.2903/j.efsa.2018.5102

Jeyasankar A, Premalatha S, Elumalai K (2014) Antifeedant and insecticidal activities of selected plant extracts against Epilachna beetle, *Henosepilachna vigintioctopunctata* (Coleoptera: Coccinellidae). Adv Entomol 2(1):14–19

Kabaluk JT, Goettel M, Erlandson M, Ericsson J, Duke G, Vernon R (2005) *Metarhizium anisopliae* as a biological control for wireworms and a report of some other naturally-occurring parasites. IOBC/WPRS Bull 28(2):109–115

Katakura H (1980) Classification and evolution of the phytophagous ladybirds belonging to *Henosepilachna vigintioctomaculata* complex (Coleoptera, Coccinellidae). PhD Dissertation, Hokkaido University, Japan

Kaya H, Alcazar J, Parsa S, Kroschel J (2009) Microbial control of the Andean potato weevil complex. Fruit Veg Cereal Sci Biotechnol 3(Special Issue 1):39–45

Keller S (2003) Integrated pest management of the potato tuber moth in cropping systems of different agroecological zones. In: Kroschel J (ed) Tropical Agriculture 11, Advances in Crop Research 1. Margraf Publishers, Weikersheim, Germany, p 153

Kennedy GG (1983) Effects of European corn borer (Lepidoptera: Pyralidae) damage on yields of spring-grown potatoes. J Econ Entomol 76(2):316–322

Kim JJ, Jeong G, Han JH, Lee S (2013) Biological control of aphid using fungal culture and culture filtrates of *Beauveria bassiana*. Mycobiology 41(4):221–224. https://doi.org/10.5941/MYCO.2013.41.4.221

Kogan M (1998) Integrated pest management: historical perspectives and contemporary developments. Annu Rev Entomol 43:243–270

Kroschel J (1994) Population dynamics of the potato tuber moth, *Phthorimaea operculella*, in Yemen and its effects on yield. In: Proceedings of the Brighton crop protection conference, pests and diseases, vol I. Brighton, UK, pp 241–246

Kroschel J (1995) Integrated Pest Management in potato production in the Republic of Yemen with special reference to the integrated biological control of the potato tuber moth (*Phthorimaea operculella* Zeller). Tropical Agriculture 8. Margraf Verlag, Weikersheim, Germany, p 227

Kroschel J, Koch W (1996) Studies on the use of chemicals, botanicals and *Bacillus thuringiensis* in the management of the potato tuber moth in potato stores. Crop Prot 15(2):197–203

Kroschel J, Schaub B (2013) Biology and ecology of potato tuber moths as major pests of potato. In: Giordanengo P, Vincent C, Alyokhin A (eds) Insect pests of potato: global perspectives on biology and management. Academic Press, Oxford, pp 165–192

Kroschel J, Sporleder M (2006) Ecological approaches to integrated pest management of the potato tuber moth, *Phthorimaea operculella* Zeller (Lepidoptera, Gelechidae). In: Proceedings of the 45th Annual Washington State Potato Conference, February 7–9, 2006, Moses Lake, Washington, pp 85–94

Kroschel J, Zegarra O (2010) Attract-and-kill: a new strategy for the management of the potato tuber moths *Phthorimaea operculella* (Zeller) and *Symmetrischema tangolias* (Gyen) in potato—Laboratory experiments towards optimizing pheromone and insecticide concentration. Pest Manage Sci 66:490–496

Kroschel J, Zegarra O (2013) Attract-and-kill as a new strategy for the management of the potato tuber moths *Phthorimaea operculella* (Zeller) and *Symmetrischema tangolias* (Gyen) in potato: evaluation of its efficacy under potato field and storage conditions. Pest Manage Sci 69(11):1205–1215

Kroschel J, Kaack HJ, Fritsch E, Huber J (1996) Biological control of the potato tuber moth (*Phthorimaea operculella* Zeller) in the Republic of Yemen using granulosis virus: Propagation and efficacy of the virus in field trials. Biocontrol Sci Technol 6:217–226

Kroschel J, Alcazar J, Pomar P (2009) Potential of plastic barriers to control Andean potato weevil *Premnotrypes suturicallus* Kuschel. Crop Prot 28:466–476

Kroschel J, Cañedo V, Alcázar J (2011) Manual de capacitación para extensionistas: Manejo de plagas de la papa en la región Andina del Perú. International Potato Center, Lima, Peru, p 88

Kroschel J, Mujica N, Alcazar J, Canedo V, Zegarra O (2012) Developing integrated pest management for potato: experiences and lessons from two distinct potato production systems of Peru. In: He Z, Larkin RP, Honeycutt CW (eds) Sustainable potato production: global case studies. Springer, UK, pp 419–450

Kroschel J, Sporleder M, Tonnang HEZ, Juarez H, Carhuapoma P, Gonzales JC, Simon R (2013) Predicting climate change caused changes in global temperature on potato tuber moth *Phthorimaea operculella* (Zeller) distribution and abundance using phenology modeling and GIS mapping. Agric For Meteorol 170:228–241

Kroschel J, Mujica N, Carhuapoma P, Sporleder M (2016a) Pest Distribution and Risk Atlas for Africa—potential global and regional distribution and abundance of agricultural and horticultural pests and associated biocontrol agents under current and future climates. International Potato Center (CIP), Lima, Peru, p 650. https://doi.org/10.4160/9789290604761. (online publication: http://cipotato.org/riskatlasforafrica/)

Kroschel J, Sporleder M, Carhuapoma P (2016b) Potato tuber moth, *Phthorimaea operculella* (Zeller). In: Kroschel J, Mujica N, Carhuapoma P, Sporleder M (eds) Pest distribution and risk atlas for Africa—potential global and regional distribution and abundance of agricultural and horticultural pests and associated biocontrol agents under current and future climates. International Potato Center, Lima, Peru, pp 7–23. https://doi.org/10.4160/9789290604761. (online publication: http://cipotato.org/riskatlasforafrica/)

Lacey L, Kroschel J (2009) Microbial control of the potato tuber moth (Lepidoptera: Gelechiidae). Fruit, Veg Cereal Sci Biotech 3:46–54

Lacey LA, Liu T-X, Buchman JL, Munyaneza JE, Goolsby JA, Horton DR (2011) Entomopathogenic fungi (Hypocreales) for control of potato psyllid, *Bactericera cockerelli* (Šulc) (Hemiptera: Triozidae) in an area endemic for zebra chip disease of potato. Biol Control 56 (3):271–278

Landis BJ, Davis EW (1947) Two-spotted spider mite damage to potatoes. J Econ Entomol 40(4):565–565

Lara FM, Scaranello AL, Baldin ELL, Boiça-Junior L, Lourenção AL (2004) Resistência de genótipos de batata a larvas e adultos de *Diabrotica speciosa*. Hortic Bras 22(4):761–765

Larrain P, Kalazich J, Carrillo R, Cisternas E (2003) Plagas de la papa y su manejo. Colección libros INIA- N° 9. Instituto de Investigaciones Agropecuarias, Ministerio de Agricultura. La Serena, Chile

Laznik Ž, Tóth T, Lakatos T, Vidrih M, Trdan S (2010) Control of the Colorado potato beetle (*Leptinotarsa decemlineata* [Say]) on potato under field conditions: a comparison of the efficacy of foliar application of two strains of *Steinernema feltiae* (Filipjev) and spraying with thiametoxam. J Plant Dis Prot 117(3):129–135

Learmonth S (2017) Thrips: potato pest in Indonesia and Western Australia. Department of Primary Industries and Regional Development. https://www.agric.wa.gov.au/potatoes/thrips-potato-pest-indonesia-and-western-australia

Li WD, Chen SX, Qin JG (2003) Identification of sympatric Asian and European corn borers. Entomol Knowl 40(1):31–35

Malumphy C, Everatt M, Eyre D, Giltrap N (2016) Potato flea beetle, *Epitrix* species. Plant pest factsheet. Department for Environment Food & Rural Affairs, United Kingdom, pp 1–6

Mamani D, Sporleder M, Kroschel J (2011) Efecto de materiales inertes de fórmulas bioinsecticidas en la protección de tubérculos almacenados contra las polillas de papa (Effect of inert materials of bioinsecticides formula to protect stored tubers of potato against the potato moths). Rev Peru Entomol 46:43–49

Martin NA (2016) Tomato potato psyllid—*Bactericera cockerelli*. Interesting insects and other invertebrates. New Zealand Arthropod Factsheet Series Number 60. http://nzacfactsheets.landcareresearch.co.nz/Index.html

Mauchline N, Stannard K (2013) Evaluation of selected entomopathogenic fungi and bioinsecticides against *Bactericera cockerelli* (Hemiptera). NZ Plant Prot 66:324-332. https://doi.org/10.30843/nzpp.2013.66.5707

Medeiros A, Tingey W, De Jong W (2004) Mechanisms of resistance to potato leafhopper, *Empoasca fabae* (Harris). Am J Potato Res 81(6):431–441

Misra SS (1995) White grub, *Holotrichia* (*Lachnosterna*) *coracea* (Hope)-A key pest of potatoes in Himachal Pradesh (India). J Entomol Res 9(2):181–182

Misra SS, Chandel RS (2003) Potato white grubs in India. Tech Bull 60

Muchemi SK, Ngichop FC, Fiaboe KKM (2014) Report of the establishmnet of the released hymenoptera *Phaedrotoma scabriventris* Nixon (Braconidae), Halticpotera arduine (Walker) (Pteromalidae) and Chrysocharis flacilla (Walker) (Eulophidae) in vegetables production systems of Kenya. Submitted to Kenyan Plant Health Inspectorate Service (KEPHIS). International Centre of Insect Physiology and Ecology (icipe), 13 pp.

Mujica N (1998) Uso de trampas amarillas en el control de adultos de mosca minadora y mosca blanca. International Potato Center, Lima, Peru. Hoja divulgativa N° 1

Mujica N (2016) Ecological approaches to manage the leafminer fly *Liriomyza huidobrensis* (Blanchard) in potato-based agroecosystems of Peru. In: Kroschel J (ed) Tropical agriculture 21–Advances in crop research, vol 11. Margraf Verlag, Weikersheim, Germany, p 255

Mujica N, Kroschel J (2013) Pest intensity-crop loss relationships for the leafminer fly *Liriomyza huidobrensis* (Blanchard) in different potato (*Solanum tuberosum* L.) varieties. Crop Prot 47:6–16

Mujica N, Kroschel J (2018) Ecological, economic and environmental assessments of Integrated Pest Management technologies: case study for potato lowland systems in the Cañete valley. Food Energy Secur 8:e00153. https://doi.org/10.1002/fes3.153

Mujica N, Alcázar J, Kroschel J (2013) Interacción del nematode entomopatogénico *Heterorhabditis indica* (RHABDITIDA: HETERORHABDITIDAE) y el ectoparasitoide *Diglyphus begini* (HYMENOPTERA: EULOPHIDAE) en el control de la mosca minadora *Liriomyza huidobrensis* (DIPTERA: AGROMYZIDAE). LV Convencion Nacional de Entomologia November 4–7, 2013, Lima-Peru

Mujica N, Carhuapoma P, Kroschel J (2016) Serpentine leafminer fly, *Liriomyza huidobrensis* (Blanchard 1926). In: Kroschel J, Mujica N, Carhuapoma P, Sporleder M (eds) Pest distribution and risk atlas for Africa—potential global and regional distribution and abundance of agricultural and horticultural pests and associated biocontrol agents under current and future climates. International Potato Center (CIP), Lima, Peru, pp 114–125. https://doi.org/10.4160/9789290604761 (online publication: http://cipotato.org/riskatlasforafrica/)

Mujica N, Sporleder M, Carhuapoma P, Kroschel J (2017) A temperature-dependent phenology model for the leafminer fly *Liriomyza huidobrensis* (Blanchard) (Diptera: Agromyzidae). J Econ Entomol 110(3):1333–1344

Mujica N, Fonseca C, Suarez F, Fabian F, Marchena M, Cisneros F (2000) Reducción del uso de insecticidas en el control de la mosca minadora, *Liriomyza huidobrensis* Blanchard, por medio del uso de técnicas etológicas. In: Arning G, A Lizárraga (eds.), Control Etológico. Uso de feromonas, trampas de colores y luz para el control de plagas en la agricultura sostenible. Lima, Perú

Munyaneza JE, Crosslin JM, Upton JE (2007) Association of *Bactericera cockerelli* (Homoptera: Psyllidae) with "zebra chip", a new potato disease in southwestern United State and Mexico. J Econ Entomol 100(3):656–663

Munyaneza JE, Sengoda VG, Crosslin JM, De la Rosa-Lozano G, Sánchez A (2009) First report of "*Candidatus Liberibacter psyllaurous*" in potato tubers with zebra chip disease in Mexico. Plant Dis 93(5):552

Nault BA, Kennedy GG (1996) Timing insecticide applications for managing European corn borer (Lepidoptera: Pyralidae) infestations in potato. Crop Prot 15(5):465–471

Naz F, Inayatullah M, Rafi MA, Ashfaque M, Ali A (2012) *Henosepilachna vigintioctopunctata* (Fab.) (Epilachninae; Coccinellidae); its taxonomy, distribution and host plants in Pakistan. Sarhad J Agric 28(3):421–427

Niederwieser F (2017) Guidelines for the control of tomato leafminer (*Tuta absoluta*) on potato. CHIPS 31(1):30–32

Niño L (2004) Revisión sobre la polilla de la papa *Tecia solanivora* en Centro y Suramerica. Suplemento de la Revista Latinoamericana de la papa, pp 4–22

Noetzel DM, Cutkomp LK, Harein PK (1985) Estimated annual losses due to insects in Minnesota 1981–1983. University of Minnesota Agricultural Experiment Station Bulletin 2541

Notz AP (1992) Distribution of eggs and larvae of *Scrobipalpula absoluta* in potato plants. Rev Facul Agron (Maracay) 18:425–432

Oerke EC, Dehne HW, Schönbeck F, Weber A (1994) Crop production and crop protection. Estimated losses in major food and cash crops. Elsevier Science, Amsterdam, p 808

OIRSA (2015a) El psílido de la papa y tomate *Bactericera* (*Paratrioza*) *cockerelli* (Sulc) (*Hemiptera: Triozidae*): ciclo biológico; la relación con las enfermedades de las plantas y la estrategia del manejo integrado de plagas en la región del OIRSA. https://www.oirsa.org/contenido/Manual%20Bactericera%20Cockerelli%20version%201.3.pdf

OIRSA (2015b) Manual de procedimientos para la vigilancia, prevención y control de la polilla del tomate *Tuta absoluta* (*Lepidóptera: Gelechiidae*) en la región del OIRSA. https://www.oirsa.org/contenido/2018/Sanidad_Vegetal/Manuales%20OIRSA%202015-2018/Manual%20T%C3%A9cnico%20de%20Tuta%20absoluta%20marzo%2010,%202017.pdf. Accessed 10 Oct 2017

OIRSA (2016) El psílido de la papa y tomate *Bactericera (Paratrioza) cockerelli* (Sulc) *(Hemiptera: Triozidae):* ciclo biológico; la relación con las enfermedades de las plantas y la estrategia del manejo integrado de plagas en la región del OIRSA. Editorial Tauro, Mexico. p. 48

Orlandini S, Magarey RD, Woo Park E, Sporleder M, Kroschel J (2017) Methods of agroclimatology: modeling approaches for pests and diseases. In: Hatfield JL, MVK S, Prueger JH (eds) Agroclimatology. Linking agriculture to climate. Agron Monogr 60. ASA, CSSA, and SSSA, Madison, WI. https://doi.org/10.2134/agronmonogr60.2016.0027

Orozco FA, Cole DC, Forbes G, Kroschel J, Wanigaratne S, Arica D (2009) Monitoring adherence to the International Code of Conduct—highly hazardous pesticides in central Andean agriculture and farmers' rights to health. Int J Occup Environ Health 15(3):255–268

Ortiz O, Alcázar J, Catalán W, Cerna V (1996) Economic impact of IPM practices on the Andean potato weevil in Peru. In: Walker TS, Crissman CC (eds) Case studies of the economic impact of CIP related technologies. International Potato Center (CIP), Lima, Peru, pp 95–113

Pacheco DM (2015) Patogenicidad de dos nematodos entomopatógenos (*Heterorhabditis* sp y *Steinernema* sp) sobre larvas y pupas de *Prodiplosis longifila* en cultivos de tomate (*Lycopersicum esculentum* Miller). Dissertation, Universidad Nacional Pedro Ruiz Gallo

Parker WE, Howard JJ (2001) The biology and management of wireworms (*Agriotes* spp.) on potato with particular reference to the UK. Agric For Entomol 3(2):85–98

Pathania M, Chandel RS (2016) Life history strategy and behaviour of white grub, *Brahmina coriacea* (Hope) (Coleoptera: Scarabaeidae: Melolonthinae) an invasive pest of potato and apple agro-ecosystem in northwestern India. Oriental Insects 51(1):46–69

Pell JK, Eilenberg J, Hajek AE, Steinkraus DS (2001) Biology, ecology and pest management potential of entomophthorales. In: Butt TM, Jackson C, Magan N (eds) Fungi as biocontrol agents: progress, problems and potential. CAB International, Wallingford, UK, pp 71–154

Pollet A, Barragan A, Zeddam JL, Laery X (2003) *Tecia solanivora*, a serious biological invasion of potato cultures in South America. Int Pest Control 45(3):139–144

Powell W, Pell JK (2017) Biological control. In: Van Emden HF, Harrington R (eds) Aphids as crop pests, 2nd edn. CABI Publishing, Wallingford, pp 469–514

Pundt L (2007) Biological control of two-spotted spider mites. College of Agriculture, Health and Natural Resources, UConn Extension, The University of Connecticut, USA. http://ipm.uconn.edu/documents/raw2/html/664.php?display=print. Accessed 2 Nov 2018

Puttarudriah M, Krishnamurti B (1954) Problem of *Epilachna* control in Mysore: Insecticidal control found inadvisbale when natural incidence of parasites is high. Indian J Entomol 16:137–141

Quiroz R, David AR, Kroschel J, Andrade-Piedra J, Barreda C, Condori B, Mares V, Monneveux P, Perez W (2018) Impact of climate change on the potato crop and biodiversity in its center of origin. Open Agric 3:273–283. https://doi.org/10.1515/opag-2018-0029

Radcliffe EB, Johnson KB (1994) Biology and Management of Leafhoppers on Potato. In: Zehnder GW, Powelson ML, Jansson RK, Raman KV (eds) Advances in Potato Pest Biology and Management, APS Press, Saint Paul, MN, pp 71–82

Reátegui F, Wilson J, Ayquipa G, Neyra S (2009) Actividad entomopatógena de cuatro especies de hongos sobre (Diptera: Cecidomyiidae). REBIOL 29(1):1–6

Rehman M, Melgar J, Rivera C, Urbina N, Idris AM, Brown JK (2010) First report of *Candidatus Liberibacter psyllaurous* or Ca. *Liberibacter* solanacearum associated with severe foliar chlorosis, curling, and necrosis and tuber discoloration of potato plants in Honduras. Plant Dis 94:376

Salazar L (1995) Los virus de la papa y su control. Centro Internacional de la Papa (CIP), Lima, Peru

Sanchez G, Vergara C (2002) Plagas del cultivo de papa. Departamento de Entomologia y Fitopatologia, UNALM, Lima, Peru

Sarmiento J (1997) Manejo de *Prodiplosis longifila* Gagné. Perú, p 1–4

Schaub B, Kroschel J (2017) Developing a biocontrol strategy to protect stored potato tubers from infestation with potato tuber moth species. J Appl Entomol 142(1–2):78–88. https://doi.org/10.1111/jen.12426

Schaub B, Carhuapoma P, Kroschel J (2016) Guatemalan potato tuber moth, *Tecia solanivora* (Povolny 1973). In: Kroschel J, Mujica N, Carhuapoma P, Sporleder M (eds) Pest distribution and risk atlas for Africa—potential global and regional distribution and abundance of agricultural and horticultural pests and associated biocontrol agents under current and future climates. International Potato Center (CIP), Lima, Peru, pp 24–38. https://doi.org/10.4160/9789290604761. (online publication: http://cipotato.org/riskatlasforafrica/)

SENASICA (2016) Palomilla del tomate *Tuta absoluta Meyrick*. Ficha Técnica 28:71

Song GH, Wu WW, Zhao QL (2008) Infection law and the control of 28-spot ladybird. Jilin Veg 1:50

Spencer KA (1973) Agromyzidae (Diptera) of Economic Importance. The Netherlands: Junk, The Hague

Spencer KA (1990) Host specialization in the world Agromyzidae Diptera. The Netherlands: Kluwer Academic Publisher

Sporleder M (2003) The granulovirus of the potato tuber moth *Phthorimaea operculella* Zeller – characterisation and prospects for effective mass production and pest control (Dissertation). Tropical Agriculture 13, Advances in Crop Research 3. Margraf Publishers, Weikersheim, Germany, 196 pp.

Sporleder M, Kroschel J (2008) The potato tuber moth granulovirus (PoGV): use, limitations and possibilities for field applications. In: Kroschel J, Lacey L (eds) Integrated pest management of the potato tuber moth, *Phthorimaea operculella* Zeller—a potato pest of global. Tropical Agriculture 20, Advances in Crop Research importance 10. Margraf Publishers, Weikersheim, Germany, pp 49–71

Sporleder M, Lacey L (2013) Biopesticides. In: Giordanengo P, Vincent C, Alyokhin A (eds) Insect pests of potato: global perspectives on biology and management. Academic Press, Oxford, pp 463–497

Sporleder M, Kroschel J, Gutierrez Quispe MR, Lagnaoui A (2004) A temperature-based simulation model for the potato tuber worm, *Phthorimaea operculella* Zeller (Lepidoptera; Gelechiidae). Environl Entomol 33(3):477–486

Sporleder M, Tonnang HEZ, Carhuapoma P, Gonzales JC, Juarez H, Kroschel J (2013) Insect Life Cycle Modeling (ILCYM) software—a new tool for regional and global insect pest risk assessments under current and future climate change scenarios. CABI Book Publication, pp 412–427

Sporleder M (2003) The granulovirus of the potato tuber moth *Phthorimaea operculella* Zeller – characterisation and prospects for effective mass production and pest control (Dissertation). Tropical Agriculture 13, Advances in Crop Research 3. Margraf Publishers, Weikersheim, Germany, 196 pp

Sporleder M, Carhuapoma P, Kroschel J (2016) Andean potato tuber moth, *Symmetrischema tangolias* (Gyen 1913). In: Kroschel J, Mujica N, Carhuapoma P, Sporleder M (eds) Pest distribution and risk atlas for Africa—potential global and regional distribution and abundance of agricultural and horticultural pests and associated biocontrol agents under current and future climates. International Potato Center (CIP), Lima, Peru, pp 39–53. https://doi.org/10.4160/9789290604761. (online publication: http://cipotato.org/riskatlasforafrica/)

Sporleder M, Schaub B, Aldana G, Kroschel J (2017) Temperature-dependent phenology and growth potential of the Andean potato tuber moth, *Symmetrischema tangolias* (Gyen) (Lep., Gelechiidae). J Appl Entomol 141:202–218. https://doi.org/10.1111/jen.12321

Steele C (2011) Wireworm Factsheet. Growers' advice. Potato Council. https://potatoes.ahdb.org.uk/sites/default/files/publication_upload/Growers%20Advice%20-%20Wireworm%20Factsheet.pdf

Stenberg JA (2017) A conceptual framework for integrated pest management. Trends Plant Sci 22(9):759–769. https://doi.org/10.1016/j.tplants.2017.06.010

Strand L (2006) Integrated pest management for potatoes in the western United States, vol 3316. UCANR Publications

Tolman JH, McLeod DGR, Harris CR (1986) Yield losses in potatoes, onions and rutabagas in Southwestern Ontario, Canada—the case for pest control. Crop Prot 5 (4):227–237

Venkatesha MG (2006) Seasonal occurrence of *Henosepilachna vigintioctopunctata* (F.) (Coleoptera: Coccinellidae) and its parasitoid on ashwagandha in India. J Asia-Pacific Entomol 9(3):265–268

Vereijssen J, Smith GR, Weintraub PG (2018) *Bactericera cockerelli* (Hemiptera: Triozidae) and Candidatus *Liberibacter solanacearum* in Potatoes in New Zealand: Biology, Transmission, and Implications for Management. J Integr Pest Manag 9(1):13. https://doi.org/10.1093/jipm/pmy007

Verma KD, Chandla VK, Trivedi TP (1994) Present status and management of potato pests in India. Annals of Entomology 12(2):73–79

Vincent C, Alyokhin A, Giordanengo P (2013) Potatoes and their pests—setting the stage. In: Giordanengo P, Vincent C, Alyokhin A (eds) Insect pests of potato: global perspectives on biology and management. Academic Press, Oxford, pp 3–8

Vishwakarma R, Prasad PH, Ghatak SS, Mondal S (2011) Bio-efficacy of plant extracts and entomopathogenic fungi against epilachna beetle, *Henosepilachna vigintioctopunctata* (Fabricius) infesting bottle gourd. J Insect Sci 24(1):65–70

Visser D (2012) Beetles and their grubs that damage potato tubers. Technical news. ARC-Roodeplaat. http://www.potatoes.co.za/SiteResources/documents/Beetles%20and%20their%20grubs%20that%20damaged%20potato%20tubers%202012.pdf

Visser D, Stals R (2012) *Temnorhynchus coronatus* (Fabricius) (Scarabaeidae: Dynastinae), potentially a pest of sweet potato in South Africa. Afr Entomol 20(2):402–407

Weintraub PG (2001) Effects of cyromazine and abamectin on the pea leafminer *Liriomyza huidobrensis* (Diptera: *Agromyzidae*) and its parasitoid *Diglyphus isaea* (Hymenoptera: *Eulophidae*) in potatoes. Crop Prot 20:207–213

Weintraub PG, Scheffer SJ, Visser D, Valladares G, Soares Correa A, Shepard BM, Rauf A, Murphy ST, Mujica N, MacVean C, Kroschel J, Kishinevsky M, Joshi RC, Johansen NS, Hallett RH, Civelek HS, Chen B, Blanco Metzler H (2017) The invasive *Liriomyza huidobrensis* (Diptera: Agromyzidae): understanding its pest status and management globally. J Insect Sci 17(1):1–27. https://doi.org/10.1093/jisesa/iew121

Workneh F, Henne DC, Childers AC, Paetzold L, Rush CM (2012) Assessments of the Edge Effect in Intensity of Potato Zebra Chip Disease. Plant Dis 96 (7):943–947

Xu R, Zhou J, Chen Z, Jia Y, Xu K (2013) Phototactic Behavior of *Henosepilachna vigintioctomaculata* Motschulsky (Coleoptera: Coccinellidae). The Coleopterists Bulletin 69(4):806–812

Yao XL, Li CQ, Gao ZL, Lu HM, Lan L, Liu H (1992) Preliminary observations on biology of potato ladybird. J Shaanxi Agric Sci 5(1):28–29

Zhou J, Xu R, Chen Z, Jia Y, Xu K (2015) Phototactic behavior of *Henosepilachna vigintioctomaculata* Motschulsky (Coleoptera: Coccinellidae). Coleopt Bull 69(4):806–812

Zuckerman L (1999) The potato: how the humble spud rescued the western world. North Point Press, New York

Permissions

All chapters in this book were first published in TPC, by Springer Nature; hereby published with permission under the Creative Commons Attribution License or equivalent. Every chapter published in this book has been scrutinized by our experts. Their significance has been extensively debated. The topics covered herein carry significant findings which will fuel the growth of the discipline. They may even be implemented as practical applications or may be referred to as a beginning point for another development.

The contributors of this book come from diverse backgrounds, making this book a truly international effort. This book will bring forth new frontiers with its revolutionizing research information and detailed analysis of the nascent developments around the world.

We would like to thank all the contributing authors for lending their expertise to make the book truly unique. They have played a crucial role in the development of this book. Without their invaluable contributions this book wouldn't have been possible. They have made vital efforts to compile up to date information on the varied aspects of this subject to make this book a valuable addition to the collection of many professionals and students.

This book was conceptualized with the vision of imparting up-to-date information and advanced data in this field. To ensure the same, a matchless editorial board was set up. Every individual on the board went through rigorous rounds of assessment to prove their worth. After which they invested a large part of their time researching and compiling the most relevant data for our readers.

The editorial board has been involved in producing this book since its inception. They have spent rigorous hours researching and exploring the diverse topics which have resulted in the successful publishing of this book. They have passed on their knowledge of decades through this book. To expedite this challenging task, the publisher supported the team at every step. A small team of assistant editors was also appointed to further simplify the editing procedure and attain best results for the readers.

Apart from the editorial board, the designing team has also invested a significant amount of their time in understanding the subject and creating the most relevant covers. They scrutinized every image to scout for the most suitable representation of the subject and create an appropriate cover for the book.

The publishing team has been an ardent support to the editorial, designing and production team. Their endless efforts to recruit the best for this project, has resulted in the accomplishment of this book. They are a veteran in the field of academics and their pool of knowledge is as vast as their experience in printing. Their expertise and guidance has proved useful at every step. Their uncompromising quality standards have made this book an exceptional effort. Their encouragement from time to time has been an inspiration for everyone.

The publisher and the editorial board hope that this book will prove to be a valuable piece of knowledge for researchers, students, practitioners and scholars across the globe.

List of Contributors

M. W. Bonierbale
International Potato Center, Lima, Peru
Duquesa Business Centre, Manilva, Malaga, Spain

W. de Jong
Cornell University, Ithaca, NY, USA

D. Ellis, A. Salas, O. Chavez, R. J. Andrade-Piedra, H. Lindqvist-Kreuze, W. Perez Gomez, N. Anglin, W. R. Amoros and E. Salas
International Potato Center, Lima, Peru

B. Adolf and H. Hausladen
Technische Universität München, Freising, Germany

F. Bittara Molina
J. R. Simplot Company, Plant Sciences, Boise, ID, USA

J. Przetakiewicz
Plant Breeding and Acclimatization Institute, Radzikow, Blonie, Poland

P. Kromann K. Sharma and M. L. Parker
International Potato Center, Nairobi, Kenya

A. Lees
The James Hutton Institute, Invergowrie, Dundee, UK

G. A. Secor
Department of Plant Pathology, North Dakota State University, Fargo, ND, USA

A. Charkowski
Colorado State University, Fort Collins, CO, USA

J. Elphinstone
Food and Environment Research Agency (Fera) Ltd., Sand Hutton, York, UK

J. Kroschel
Independent Consultant for International Agricultural Research & Development, Filderstadt, Germany

N. Mujica
Independent Consultant, Lima, Peru

J. Okonya
International Potato Center, Kampala, Uganda

A. Alyokhin
University of Maine, Orono, ME, USA

Index